Analog Circuits and Signal Processing

Series Editors
Mohammed Ismail, Khalifa University, Dublin, OH, USA
Mohamad Sawan, Montreal, QC, Canada

The *Analog Circuits and Signal Processing* book series, formerly known as the *Kluwer International Series in Engineering and Computer Science*, is a high level academic and professional series publishing research on the design and applications of analog integrated circuits and signal processing circuits and systems. Typically per year we publish between 5–15 research monographs, professional books, handbooks, edited volumes and textbooks with worldwide distribution to engineers, researchers, educators, and libraries.

The book series promotes and expedites the dissemination of new research results and tutorial views in the analog field. There is an exciting and large volume of research activity in the field worldwide. Researchers are striving to bridge the gap between classical analog work and recent advances in very large scale integration (VLSI) technologies with improved analog capabilities. Analog VLSI has been recognized as a major technology for future information processing. Analog work is showing signs of dramatic changes with emphasis on interdisciplinary research efforts combining device/circuit/technology issues. Consequently, new design concepts, strategies and design tools are being unveiled.

Topics of interest include:
Analog Interface Circuits and Systems;
Data converters;
Active-RC, switched-capacitor and continuous-time integrated filters;
Mixed analog/digital VLSI;
Simulation and modeling, mixed-mode simulation;
Analog nonlinear and computational circuits and signal processing;
Analog Artificial Neural Networks/Artificial Intelligence;
Current-mode Signal Processing;
Computer-Aided Design (CAD) tools;
Analog Design in emerging technologies (Scalable CMOS, BiCMOS, GaAs, heterojunction and floating gate technologies, etc.);
Analog Design for Test;
Integrated sensors and actuators;
Analog Design Automation/Knowledge-based Systems;
Analog VLSI cell libraries;
Analog product development;
RF Front ends, Wireless communications and Microwave Circuits;
Analog behavioral modeling, Analog HDL.

More information about this series at http://www.springer.com/series/7381

Dawei Mai • Michael Peter Kennedy

Wandering Spurs in MASH-Based Fractional-N Frequency Synthesizers

How They Arise and How to Get Rid of Them

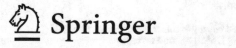 Springer

Dawei Mai
University College Dublin
Dublin, Ireland

Michael Peter Kennedy
University College Dublin
Dublin, Ireland

ISSN 1872-082X   ISSN 2197-1854   (electronic)
Analog Circuits and Signal Processing
ISBN 978-3-030-91287-1   ISBN 978-3-030-91285-7   (eBook)
https://doi.org/10.1007/978-3-030-91285-7

This Springer imprint is published by the registered company Springer Nature Switzerland AG
The registered company address is: Gewerbestrasse 11, 6330 Cham, Switzerland

To my family (D. M.)
To Rossana (M. P. K.)

Preface

Fractional-N frequency synthesizers are commonly used in communication systems. A divider controller is employed within a conventional divider-based fractional-N frequency synthesizer to achieve the desired fractional division ratio. A Multi-stAge noiSe-sHaping Digital Delta-Sigma Modulator (MASH DDSM) is usually used as the divider controller because of its high-pass shaped quantization error and ease of implementation. Among contributors to the output phase noise of a conventional fractional-N frequency synthesizer, the noise of the divider controller is normally high-pass shaped and its in-band contribution can ideally be ignored under the assumption of linearity. However, in the presence of nonlinearity, the phase noise contribution of the divider controller can become prominent in-band and spurious tones (abbreviated as "spurs") appear. The causes of divider controller-induced spurious tones in the long-term spectrum have been the focus of research for many years, and efforts continue to be made to identify and mitigate nonlinearity-induced spurs in fractional-N frequency synthesizers.

In addition to these spurs at fixed frequencies in the long-term spectrum (so-called "fixed" spurs), the short-term spectrum of a MASH DDSM-based fractional-N synthesizer can exhibit a type of time-varying spur phenomenon that is called wandering spurs. This type of spur appears in the short-term spectrum of the fractional-N frequency synthesizer and can impact applications such as radar and burst communications that depend on short windows of operation. Observations and analysis suggest that the phenomenon of wandering spurs originates in the MASH DDSM divider controller. In the third-order MASH 1-1-1 divider controller-based synthesizer considered in this book, the wandering spurs can be categorized based on the input of the MASH DDSM. Each type of wandering spur has a distinct cause.

Techniques to mitigate wandering spurs have been proposed for MASH-based synthesizers. In this book, dither-based solutions built on the commonly used MASH 1-1-1 DDSM and dither-independent solutions that exploit modifications of the original MASH 1-1-1 architecture are analyzed and compared.

The book is structured as follows:

In Chap. 2, a conventional fractional-N frequency synthesizer and simulation methods to study it are reviewed.

In Chap. 3, divider controller-induced fixed spurs in a MASH DDSM-based fractional-N frequency synthesizer are described in detail.

In Chap. 4, the root cause of wandering spurs is revealed.

In Chap. 5, analysis of the three distinct cases of wandering spurs is presented.

In Chap. 6, techniques to mitigate wandering spurs are developed based on the knowledge of their causes.

In Chap. 7, measurement results for an implemented fractional-N frequency synthesizer demonstrator with MASH-based divider controllers for wandering spur mitigation are presented.

After completing this book, we hope that the reader will have both an intuitive and a mathematical understanding of the phenomenon of wandering spurs. The reader will know how they arise and how to eliminate or mitigate them.

Dublin, Ireland
August 2021

Dawei Mai
M. Peter Kennedy

Acknowledgments

We would like to thank Yann Donnelly from UCC/Tyndall National Institute, Stefano Tulisi, James Breslin, Pat Griffin, Michael Connor, Stephen Brookes, Brian Shelly, and Michael Keaveney from the team at Analog Devices Limerick for their excellent work in implementing and testing the demonstrator synthesizer in this book.

The screenshots of real-time spectra, phase noise spectra, and spectra from Fig. 7.6 through Fig. 7.27 are owned by Analog Devices, Inc. ("ADI"), copyright © 2021. All Rights Reserved. These images, icons, and marks are reproduced with permission by ADI. No unauthorized reproduction, distribution, or usage is permitted without ADI's written consent.

Dublin, Ireland
August 2021

Dawei Mai
M. Peter Kennedy

Contents

Acronyms

AC	Alternating Current; also, time-varying
BiCMOS	Bipolar Complementary Metal Oxide Semiconductor
BW	BandWidth
CP	Charge Pump
CTFT	Continuous-Time Fourier Transform
DC	Direct Current; also, constant
DDSM	Digital Delta-Sigma Modulator
DSM	Delta-Sigma Modulator
DTFT	Discrete-Time Fourier Transform
EFM	Error Feedback Modulator
FBS	Fractional Boundary Spur
FF	Flip-Flop
IBS	Integer Boundary Spur
LCM	Least Common Multiple
LFSR	Linear Feedback Shift Register
LO	Local Oscillator
LSB	Least Significant Bit
MASH	Multi-stAge noiSe sHaping (Architecture)
MMD	Multi-Modulus Divider
NL	Nonlinear; Nonlinearity
NTF	Noise Transfer Function
PCB	Printed Circuit Board
PD	Phase Detector
PFD	Phase/Frequency Detector
PLL	Phase-Locked Loop
PSD	Power Spectral Density
PWL	PieceWise-Linear
RADAR	RAdio Detection And Ranging
RBW	Resolution BandWidth
RC	Resistor Capacitor
RF	Radio Frequency

RMS	Root Mean Square
RNG	Random Number Generator
SNR	Signal-to-Noise Ratio
SPI	Serial Peripheral Interface
STF	Signal Transfer Function
VCO	Voltage-Controlled Oscillator
WR	Wander Rate
WS	Wandering Spur
XOR	Exclusive OR

Chapter 1
Introduction

1.1 Spurious Tones in Fractional-N Frequency Synthesizers

1.1.1 Nonlinearity-Induced Fixed Spurious Tones

Frequency synthesizers are used in a wide range of electronic systems for communications and clocking purposes. Their primary function is to produce a signal (typically a sinewave or a squarewave) at a precise frequency.

The spectrum of the output signal should ideally be a single tone in the case of a sinewave, as shown in Fig. 1.1a and b, or a fundamental tone plus integer multiples thereof (harmonics) when the signal is a squarewave.

In practice, the synthesizer's output signal is noisy, leading to blurring of the otherwise pure tones in the frequency spectrum, as shown in Fig. 1.1c and d. Noise close to the fundamental frequency of the desired signal is characterized by its phase noise spectrum and/or jitter.

In this book, we consider an indirect frequency synthesizer architecture based on a phase-locked loop (PLL). A generic charge-pump (CP) PLL-based frequency synthesizer is shown schematically in Fig. 1.2. Variants of this basic architecture may include prescalers and digital to time converters.

A feedback loop is used in order to generate a precise output frequency by phase locking a controllable oscillator to an accurate reference clock. An integer-N frequency synthesizer has a divider with a *fixed* division ratio, i.e. no control to the division ratio is required. One major limitation of having a fixed division ratio is the frequency resolution at the output. Since, in steady-state operation,

$$f_{OUT} = N_{int} f_{PFD},\qquad(1.1)$$

where f_{OUT} and f_{PFD} are the output frequency and the reference input at the phase/frequency detector (PFD) and N_{int} is the integer division ratio, the output

© The Author(s), under exclusive license to Springer Nature Switzerland AG 2022
D. Mai, M. P. Kennedy, *Wandering Spurs in MASH-based Fractional-N Frequency Synthesizers*, Analog Circuits and Signal Processing,
https://doi.org/10.1007/978-3-030-91285-7_1

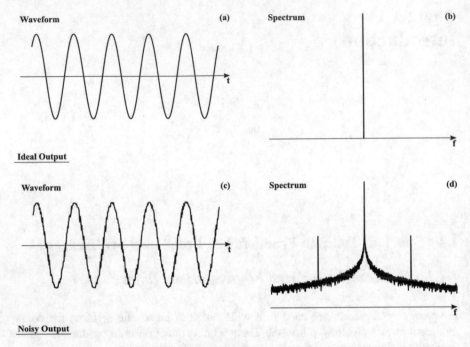

Fig. 1.1 Illustration of waveforms and spectra of an ideal (top) and noisy (bottom) sinusoidal frequency synthesizer output. Note the noise skirt and spurs in (**d**)

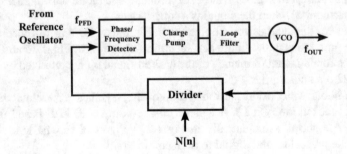

Fig. 1.2 Block diagram of a generic CP frequency synthesizer

frequency can be set with a minimum frequency resolution equal to the input reference frequency f_{PFD}. This is not sufficient in many application scenarios where it is required to have finer frequency resolution at the output.

The fractional-N frequency synthesizer provides a solution to enable a fractional ratio between the output and the input reference frequencies [1, 2]. By changing the instantaneous division ratio $N[n]$ using a divider controller, a fractional division ratio can be achieved on average. For a local oscillator application, the fixed output frequency can be expressed by

$$f_{OUT} = (N_{int} + \alpha) f_{PFD} = \left(N_{int} + \frac{X}{M} \right) f_{PFD}, \tag{1.2}$$

where α is the desired fractional part of the division ratio, X is the input to the divider controller which controls the division ratio, and M is the modulus of the divider controller. In a digital implementation, typically, $M = 2^B$ for a fixed-bus width B-bit divider controller and this is assumed throughout this book.

Despite the fact that the long-term average of the division ratio can be precise, the change of the instantaneous division ratio will introduce noise. To minimize the impact on the in-band noise, it is preferred that the additional quantization noise introduced by the divider controller should be high-pass shaped; the highpass content is subsequently removed by the lowpass transfer function from the point of injection of the output phase noise. A Multi-stAge noiSe sHaping digital delta-sigma modulator (MASH DDSM) is conventionally used for this purpose.

In a realistic fractional-N frequency synthesizer, nonlinearity exists and nonlinearity-induced excess noise and extra spurious tones at constant offsets from the nominal output frequency will appear. Figure 1.3 shows a typical phase noise measurement from a commercial fractional-N frequency synthesizer [3]. Notice the significant spurious tones, none of which is predicted by a linear model of the synthesizer [4]. These tones are typically deleterious. In a local oscillator application, for example, spurious tones will deteriorate the system performance due to the noise overlapping from the unwanted conversion they cause.

Figure 1.4 shows how the presence of spurs in a local oscillator degrade the signal-to-noise ratio (SNR) of a signal during upconversion and downconversion. Each spur produces an unwanted copy of the signal of interest. It is difficult to filter out unwanted spectral content that is introduced by spurs close to the carrier. Such close-in spurs are common in fractional-N frequency synthesis. Consequently, there has been much research investigating the mechanisms that cause *fixed* spurs in fractional-N frequency synthesizers [5–7] and how to reduce or even eliminate them under given conditions [8–11].

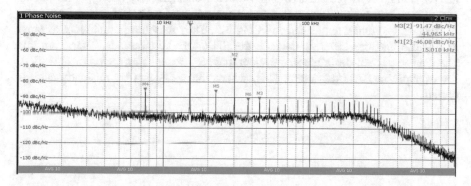

Fig. 1.3 Measured output phase noise spectrum with spurious tones (marked with blue markers) from a commercial fractional-N frequency synthesizer

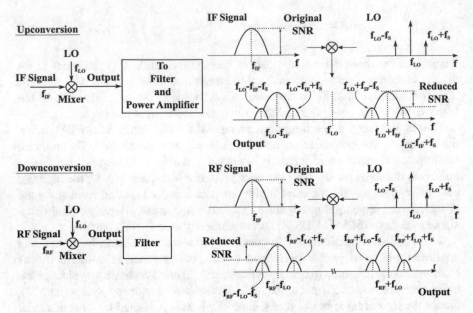

Fig. 1.4 Illustration of the degradation of SNR due to spurs during upconversion and downconversion

1.1.2   Wandering Spurs

A DDSM is conventionally used as the divider controller in a fractional-N frequency synthesizer. In other applications of Delta-Sigma Modulators (DSMs), especially audio applications [12], a time-varying frequency domain phenomenon associated with short limit cycles has been observed.

A related phenomenon of time-varying spurs, termed *wandering spurs*, has been reported in measurements and simulations of fractional-N frequency synthesizers for RF applications [13]. These spurs move with a linearly changing frequency across the short-term output phase noise spectrum of a DDSM-based fractional-N frequency synthesizer, as shown in Fig. 1.5. They may have amplitudes as large as -30 dBc and can severely degrade the SNR for short periods. Therefore, this phenomenon is not desirable in a system that requires a short period of operation, for example, a RADAR system [14], where the spurs can be misinterpreted as targets.

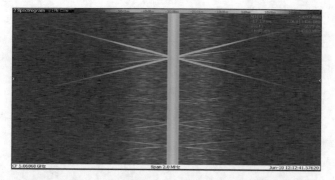

Fig. 1.5 Measured output phase noise spectrogram from commercial fractional-N frequency synthesizers. The X-shaped traces are caused by the wandering spurs

1.2 Contributions of This Book

This book has three major purposes. The first objective is to elaborate the generating mechanism that gives rise to wandering spurs. The cause of this spectral phenomenon in a conventional fractional-N frequency synthesizer is identified. The patterns of wandering spurs in a synthesizer based on a commonly used third-order divider controller are analyzed in detail.

The second purpose is to describe divider controllers for the mitigation of wandering spurs in a conventional fractional-N frequency synthesizer, on the basis of the results of theoretical analysis.

The third purpose is to present implementation details and measured results of divider controller architectures for wandering spur mitigation in a state-or-art fractional-N frequency synthesizer demonstrator to determine the effectiveness of the theoretical designs in practice.

The primary contributions of this book are as follows:

- Analysis of the cause of wandering spurs in a MASH 1-1-1 DDSM-based conventional charge-pump fractional-N frequency synthesizer;
- Categorization of the wandering spurs originating from a MASH 1-1-1 DDSM divider controller;
- Analysis of the wandering spur patterns in a MASH 1-1-1 DDSM-based conventional charge-pump fractional-N frequency synthesizer;
- Analysis of spur mitigation techniques for MASH-based divider controllers based on additive high amplitude dither;
- Analysis of dither-independent modified MASH divider controllers for wandering spur mitigation; and
- Presentation and comparison of measurement results for wandering spur mitigation techniques in a fractional-N frequency synthesizer implemented in a 180 nm SiGe BiCMOS process.

Chapter 2
Simulation of Phase Noise in a Fractional-N Frequency Synthesizer

In this chapter, the fractional-N frequency synthesizer considered and the models of it used in the simulations in this book are elaborated. Nonlinearity will cause extra noise components in the synthesizer's output, especially in the contribution from the divider controller. Simplified models characterizing the dominant nonlinearity of a fractional-N frequency synthesizer are detailed. Examples of the evaluation of phase noise performance via simulation are also presented.

2.1 Conventional Fractional-N Synthesizer

A general charge-pump fractional-N frequency synthesizer has the structure shown in Fig. 2.1. It employs a multi-modulus divider which can implement different integer division ratios. To achieve the desired fractional division ratio, control is administered by a divider controller which changes the instantaneous division ratio. During steady-state operation, the output of the synthesizer can be expressed by

$$f_{OUT} = (N_{int} + \mathbf{E}\left(y[n]\right)) f_{PFD} = \left(N_{int} + \frac{\mathbf{E}\left(x[n]\right)}{M}\right) f_{PFD}, \qquad (2.1)$$

where N_{int} is the integer part of the division ratio, X is the input to the divider controller, M is the modulus of the divider controller, n is the sample index, and $\mathbf{E}(\cdot)$ denotes expectation. The modulus M is assumed to be an integer power of two throughout this book. The presented theory can be extended to cases of arbitrary values of M. For the local oscillator application considered in this book, it is assumed the input to the divider controller is kept constant, i.e.

$$x[n] = X. \qquad (2.2)$$

© The Author(s), under exclusive license to Springer Nature Switzerland AG 2022
D. Mai, M. P. Kennedy, *Wandering Spurs in MASH-based Fractional-N Frequency Synthesizers*, Analog Circuits and Signal Processing,
https://doi.org/10.1007/978-3-030-91285-7_2

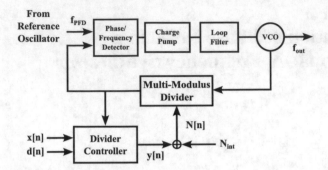

Fig. 2.1 Block diagram of a charge-pump fractional-N frequency synthesizer. The input $x[n]$ is usually a constant in the case of a local oscillator; a one-bit additive dither signal $d[n]$ is typically applied [15]

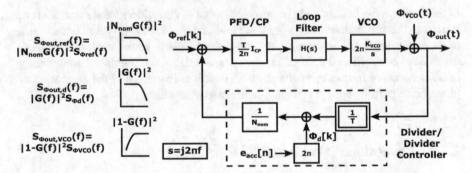

Fig. 2.2 A simplified linear phase domain model of a fractional-N frequency synthesizer with its building blocks in the s-domain [4]. $\Phi_{VCO}(t)$, $\Phi_{ref}[k]$, and $\Phi_{out}[k]$ are the phase of VCO, reference, and output, respectively. $G(f)$ is the closed-loop transfer function of the synthesizer

To analyze the phase noise of the synthesizer, a frequency domain model of the phase excursion shown in Fig. 2.2 can be used [4]. It is sensible to refer noise sources to appropriate points of injection. The output phase noise in a synthesizer can be quantified as the total of the contributions from the equivalent noise sources injected at those points. The voltage-controlled oscillator (VCO) noise, which is commonly modeled by a $1/f^2$ noise source, is an intrinsic noise source of the synthesizer. The transfer function from this noise source to the output phase noise has a high-pass transfer characteristic. In this book, reference path noise refers to the combined equivalent for noise sources from various parts of the synthesizer. For example, the combined equivalent for reference jitter and charge pump noise experiences a low-pass transfer function when contributing to the output phase noise. The equivalent reference input-referred noise is commonly modeled by a white noise source [4].

As mentioned in Chap. 1, the noise generated by the instantaneous change of the division ratio, namely the quantization error of the divider controller, contributes to the output phase noise [4]. Since the transfer function between this source of

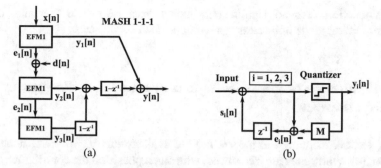

Fig. 2.3 Block diagrams of (**a**) a MASH 1-1-1 and (**b**) a first-order error feedback modulator (EFM1). For a conventional LSB-dithered MASH 1-1-1, a 1-bit dither $d[n]$ is introduced to reduce the periodicity in the quantization error

Fig. 2.4 Contributions to the output phase noise spectrum of a typical fractional-N synthesizer with a MASH 1-1-1 based divider controller based on a linearized model [4]

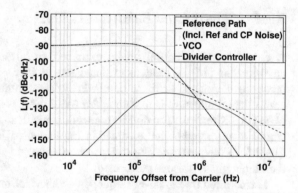

noise and the output phase noise has a low-pass characteristic, high-pass shaping this quantization noise can benefit the overall phase noise performance.

A MASH DDSM is commonly used as the divider controller due to its high-pass shaped quantization noise, guaranteed stability, and ease of implementation. The first-order MASH is a simple digital accumulator and its quantization error is periodic and prone to exhibiting spectral tones. Quantization noise of a second-order or higher MASH can provide sufficiently randomized quantization error under appropriate conditions. A third-order MASH DDSM, which is shown schematically in Fig. 2.3, is widely used because of its relatively ideal quantization noise, which is close to shaped white noise [1].

In a typical fractional-N frequency synthesizer, the major noise contributors are the reference path noise (which includes reference phase noise and charge pump noise), VCO intrinsic noise, and the noise from the divider controller, as shown in Fig. 2.4 [4].

Analysis using a simplified linearized phase domain model of the synthesizer shows that the input reference path noise dominates the in-band phase noise performance and that the VCO noise is dominant outside the loop bandwidth. The high-pass shaped divider controller noise, which experiences the low-pass loop transfer function, does not significantly impact the noise profile over most of the frequency offset range of interest in a well-designed synthesizer.

2.2 Evaluation of Phase Noise in a Fractional-N Frequency Synthesizer

In this section, methods and models used to evaluate the output phase noise of a fractional-N synthesizer are introduced, with an emphasis on the contribution of the divider controller in the presence of nonlinearity.

2.2.1 Closed-Loop Behavioral Model

A closed-loop behavioral model extracted from the CppSim System Simulator environment [16] is used to evaluate the divider controller performance in the presence of a piecewise linear nonlinearity. The synthesizer loop without the divider controller architecture is first completed and used as an 's-function' module within MATLAB Simulink. The divider controller, which applies an integer-valued control signal to the synthesizer module, is implemented in Simulink, as illustrated in Fig. 2.5. In this way, the closed-loop simulation of waveforms at critical nodes and the output phase noise is performed.

The detailed fractional-N frequency synthesizer model is shown in Fig. 2.6. The set of parameters of the model synthesizer used in this book is given in Table 2.1.

Note that an up/down current mismatch ϵ of 8% is considered throughout this book, unless otherwise stated. The 8% mismatch of the charge pump currents is used because it leads to prominent divider controller-induced spurious tones. A reset

Fig. 2.5 CppSim-based closed-loop simulation

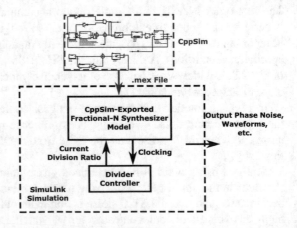

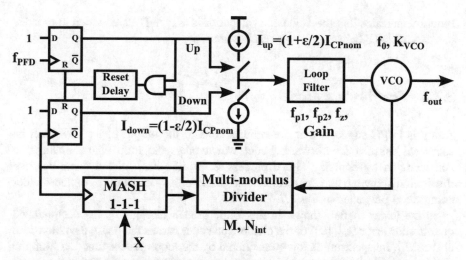

Fig. 2.6 Block diagram of the simulated fractional-*N* frequency synthesizer with a charge pump up and down current mismatch of ϵ

Table 2.1 Model parameters used in the simulations

Parameter	Value
Simulation frequency f_s	2 GHz
f_{PFD}	20 MHz
PFD reset delay	0.5 ns
I_{CPnom}	1.5 mA
Closed-loop bandwidth	150 kHz
Gain K_H	229.14×10^6
f_{p1}	663 kHz
f_{p2}	300 kHz
f_z	18.75 kHz
f_0	900 MHz
K_{vco}	10 MHz/V
N_{int}	45
M	2^{20}

delay, which is used to eliminate the dead zone of the detector, is also included [17, 18].

The loop filter has a transfer function of

$$H(s) = K_H \frac{1 + \frac{s}{2\pi f_z}}{s(1 + \frac{s}{2\pi f_{p1}})(1 + \frac{s}{2\pi f_{p2}})}. \tag{2.3}$$

The zero, poles, and gain are chosen to give a bandwidth of the synthesizer that is less than one-tenth of the reference frequency and yields a relatively flat in-band response, which ensures the stability of the synthesizer. The loop filter transfer function indicates that the modeled synthesizer is a Type-II PLL, which in practice has negligible DC phase offset in steady state.

2.2.2   Feedforward Model

Analysis in [4] suggests that the contributions to the output phase noise can be estimated using a feedforward model which takes the loop characteristic into consideration by applying signal processing to the phase-related quantities. An illustration showing how to estimate the phase noise contribution of the divider controller is presented in Fig. 2.7.

In the linear model, shown in the *lower* branch in Fig. 2.7, the accumulated quantization error ($e_{acc}[n]$) of the divider controller causes the phase deviation seen at the PFD. Its spectrum is low-pass filtered by the loop transfer function from the divider controller to the output of the synthesizer $G(f)$, giving the *linear prediction* of the phase noise component [4].

The effect of a typical nonlinear characteristic of the phase/frequency detection and the charge pump in a synthesizer can be taken into account in the feedforward model, as shown in the *upper* branch in Fig. 2.7 [19]. The corresponding distorted version of $e_{acc}[n]$ in the presence of a memoryless nonlinearity,[1] denoted $e_{acc}^{NL}[n]$, is shaped by the loop and leads to a nonlinear contribution to the output phase noise. In this case, the output spectrum of the contribution from the divider controller in the synthesizer shown in Fig. 2.6 can be expressed by

$$S_{out}(f) = S_{e_{acc}^{NL}}(f) \left| \frac{2\pi A(f)}{1 + A(f)} \right|^2 \tag{2.4}$$

$$= S_{e_{acc}^{NL}}(f) \left| 2\pi G(f) \right|^2, \tag{2.5}$$

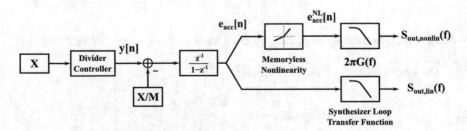

Fig. 2.7 Block diagram representation of the feedforward model [15]

---

[1] In this book, only memoryless nonlinearities are considered.

where

$$A(f) = \frac{1}{2\pi N_{nom}} I_{CP} H(f) \frac{K_{vco}}{jf} \tag{2.6}$$

is the open-loop transfer function, $N_{nom} = N_{int} + X/M$ is the nominal division ratio, I_{CP} is the nominal charge pump current, $H(f)$ is the loop filter transfer function and K_{vco} is the gain of the VCO. Note that this is an *approximate* model and it is valid for weak memoryless nonlinearities which are typically found in conventional charge-pump synthesizers.

The feedforward model provides a concise method to evaluate the performance of the divider controller that needs significantly less computation time than a full closed-loop behavioral simulation. The key spur generation mechanism, namely that due to the interaction between the accumulated quantization error and the memoryless nonlinearity, is captured by the model with fidelity and can be used to compare different divider controller designs. This will be further discussed in the next section.

2.2.3   Evaluation of Phase Noise

The output of a synthesizer can be subject to both amplitude and phase deviation. In the simulations and theoretical analysis in the phase domain, it is commonly assumed and modeled that the output phase of the synthesizer has a very limited excursion while the signal is accurate in amplitude, i.e., the output amplitude is a constant. The output of the synthesizer at f_0 can be expressed as

$$x(t) = \cos(\omega_0 t + \phi(t)) \tag{2.7}$$

$$\approx \cos(\omega_0 t) - \phi(t)\sin(\omega_0 t) \tag{2.8}$$

where $\omega_0 = 2\pi f_0$.

The original definition of the phase noise at an offset frequency Δf is [20]

$$\mathcal{L}(\Delta f) = \frac{\text{power density in one phase noise modulation sideband, per Hz}}{\text{total signal power}}. \tag{2.9}$$

To evaluate the phase noise, the power spectral density of the output phase deviation is first computed. The autocorrelation of the signal is

$$R_{xx}(\tau) = \mathbf{E}(x(t)x(t+\tau)) \approx \tag{2.10}$$

$$\lim_{T\to\infty} \frac{1}{T} \int_0^T (\cos(\omega_0 t) - \phi(t)\sin(\omega_0 t))$$

$$\times (\cos(\omega_0(t+\tau)) - \phi(t+\tau)\sin(\omega_0(t+\tau))) \, dt \tag{2.11}$$

$$= \lim_{T \to \infty} \left(\frac{1}{T} \int_0^T \phi(t) \sin(\omega_0 t) \phi(t+\tau) \sin(\omega_0(t+\tau)) dt \right.$$

$$+ \frac{1}{T} \int_0^T \cos(\omega_0 t) \cos(\omega_0(t+\tau)) dt - \frac{1}{T} \int_0^T \cos(\omega_0 t) \phi(t+\tau) \sin(\omega_0(t+\tau)) dt$$

$$\left. - \frac{1}{T} \int_0^T \cos(\omega_0(t+\tau)) \phi(t) \sin(\omega_0 t) dt \right)$$

$$\tag{2.12}$$

$$\approx \frac{1}{2} \cos(\omega_0 \tau) R_{\phi\phi}(\tau) + \frac{1}{2} \cos(\omega_0 \tau), \tag{2.13}$$

where $R_{\phi\phi}(\tau) = \mathbf{E}(\phi(t)\phi(t+\tau))$. This gives a power spectral density of

$$S_x(f) = \mathcal{F}(R_{xx}(\tau)) = \int_{-\infty}^{\infty} R_{xx}(\tau) e^{-j2\pi f \tau} d\tau \tag{2.14}$$

$$\approx \frac{1}{4} \delta(f+f_0) + \frac{1}{4} \delta(f-f_0) + \frac{1}{4} S_\phi(f+f_0) + \frac{1}{4} S_\phi(f-f_0), \tag{2.15}$$

and

$$S_\phi(f) = \mathcal{F}(R_{\phi\phi}(\tau)) = \int_{-\infty}^{\infty} R_{\phi\phi}(\tau) e^{-j2\pi f t}. \tag{2.16}$$

According to the definition, the phase noise can be expressed as

$$\mathcal{L}(\Delta f) \approx \frac{2 \times \frac{1}{4} S_\phi(f-f_0)}{1/2} = S_\phi(f-f_0) = S_\phi(\Delta f). \tag{2.17}$$

In (2.17), the denominator is the power of the cosine wave, assuming it contains the majority of the signal. The numerator is the approximate amount of power in the signal per hertz at an offset frequency $\Delta f = f - f_0$. The power spectral density of the phase excursion $S_\phi(f)$ has the units of rad^2/Hz. The phase noise is the ratio between the power per hertz at an offset and the total signal power; it is measured in decibels per hertz (dB/Hz). Also, since it is referred to the carrier power, the units dBc/Hz are usually used. Since phase noise is closely related to the power spectral density $S_\phi(f)$ and can be approximated by it, following [21], the phase noise is calculated using the definition

$$\mathcal{L}(\Delta f) \equiv S_\phi(\Delta f), \tag{2.18}$$

where $S_\phi(f)$ is the double-sided power spectral density and the result has the units of dBc/Hz. The definition is valid for cases where the mean squared phase deviation is less than 0.1 rad^2 [21].

In both the closed-loop behavioral model and the feedforward model simulations, the output related to the phase deviation is described in the phase domain and the data sequence is uniformly sampled.

Let $\phi_{con}(t)$ be the continuous phase deviation. The sampled version of it with a sampling period of T_s is then

$$\phi_{con,sa}(t) = \phi_{con}(t) \sum_{n=-\infty}^{\infty} \delta(t - nT_s). \tag{2.19}$$

The Fourier transformation of this is

$$\Phi_{con,sa}(f) = \frac{1}{T_s} \sum_{n=-\infty}^{\infty} \Phi_{con}(f)\delta\left(f - \frac{k}{T_s}\right) \approx \frac{1}{T_s}\Phi_{con}(f). \tag{2.20}$$

The last approximation assumes negligible aliasing after sampling.

Also, $\Phi_{con,sa}(f)$ can be related to the discrete-time Fourier transform of the corresponding discrete phase deviation sequence $\Phi_{dis}(F)$ by

$$\Phi_{con,sa}(f) = \int_{-\infty}^{\infty} \phi_{con}(nT_s) \sum_{n=-\infty}^{\infty} \delta(t - nT_s)e^{-j2\pi fnT_s}dt \tag{2.21}$$

$$= \int_{-\infty}^{\infty} \phi_{dis}[n] \sum_{n=-\infty}^{\infty} \delta(t - nT_s)e^{-j2\pi \frac{f}{f_s}n}dt \tag{2.22}$$

$$= \sum_{n=-\infty}^{\infty} e^{-j2\pi \frac{f}{f_s}n} \int_{-\infty}^{\infty} \phi_{dis}[n]\delta(t - nT_s)dt \tag{2.23}$$

$$= \sum_{n=-\infty}^{\infty} \phi_{dis}[n]e^{-j2\pi \frac{f}{f_s}n} \tag{2.24}$$

$$= \Phi_{dis}(F), \tag{2.25}$$

where $F = f/f_s$. Comparing (2.20) and (2.25), note that

$$\Phi_{dis}(F) = \frac{1}{T_s}\Phi_{con}(f) \tag{2.26}$$

holds.

In practice, when the signal with a limited bandwidth is sampled at a sufficiently high frequency,

$$\Phi_{dis}\left(\frac{k}{N}\right) = \frac{1}{T_s}\Phi_{con}\left(\frac{k}{N}f_s\right), \tag{2.27}$$

where N is the number of samples in the sequence and integer $k \in [0, N)$, can be used to estimate the corresponding continuous spectrum from the sampled sequence.

A Fourier-based periodogram is the basis of the power spectral density estimation. The estimate is found by

$$S_{\phi_{con}}(f) = \frac{1}{T} |\Phi_{con}(f)|^2, \tag{2.28}$$

$$S_{\phi_{dis}}(F) = S_{\phi_{dis}}\left(\frac{k}{N}\right) = \frac{1}{N} \left|\Phi_{dis}\left(\frac{k}{N}\right)\right|^2, \tag{2.29}$$

where $F = k/N$ and $T = T_s N$ is the time span. Substituting (2.25) gives

$$S_{\phi_{con}}\left(\frac{k}{N} f_s\right) = \frac{1}{T} \left|T_s \Phi_{dis}\left(\frac{k}{N}\right)\right|^2 = \frac{T_s^2}{N T_s} \left|\Phi_{dis}\left(\frac{k}{N}\right)\right|^2 \tag{2.30}$$

$$= \frac{T_s}{N} \left|\Phi_{dis}\left(\frac{k}{N}\right)\right|^2. \tag{2.31}$$

By specifying the sampling frequency, MATLAB can be used to find the corresponding continuous spectral estimate by applying the scaling in (2.31).

The modified periodogram, which applies a window function $w[n]$ before the analysis, is implemented in MATLAB and the method can be expressed by [22]:

$$S(f) = \frac{T_s}{NU} \left|\sum_{n=0}^{N-1} w[n]x[n]e^{-j2\pi f T_s n}\right|^2, \quad f = \frac{k}{N} f_s, \ k \in [0, N), \tag{2.32}$$

where U is a normalization constant

$$U = \frac{1}{N} \sum_{n=0}^{N-1} |w[n]|^2. \tag{2.33}$$

When the data sequence is long, calculating the modified periodogram from sections of the discrete sequence and averaging the results can reduce the variance in the estimate of the power spectral density. Welch's method utilizes moving windows with a constant overlap to determine the power spectral density. It has an identical expected value to the modified periodogram estimate using the same window [23]. Both the modified periodogram and Welch's method are used in the evaluation of the simulation results in this book.

In the closed-loop model, the output phase sequence is available and the output phase noise can be evaluated directly. In the feedforward model, by filtering the spectrum of the noise sources associated with the reference path, VCO, and divider controller through their corresponding transfer functions to the output phase noise, as the example shown in Fig. 2.7, each contribution can be evaluated. To find the

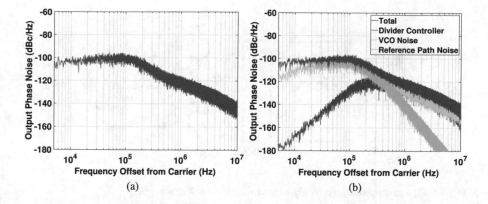

Fig. 2.8 Simulated output phase noise spectrum of a MASH 1-1-1-based *linear* fractional-N frequency synthesizer with the parameters in Table 2.1 and identical noise contributors: (**a**) closed-loop behavioral model and (**b**) feedforward model

overall phase noise, the noise sources are assumed to be independent and the three mentioned contributions are summed.

Figure 2.8a shows the closed loop simulation result for a *linear* MASH 1-1-1-based synthesizer loop with the parameters in Table 2.1. Figure 2.8b shows the feedforward simulation results for each component of the output phase noise and the overall output phase noise estimate. In both cases, a reference path noise of −137 dBc/Hz and VCO noise of −125 dBc/Hz at 1 MHz offset are implemented. The models produce similar predictions for the phase noise.

2.3 Models of Nonlinearity in Fractional-N Frequency Synthesizer

2.3.1 Spectral Impact of Nonlinearity

Unfortunately, the linear assumption does not hold in practice. Nonlinearity exists in every practical fractional-N frequency synthesizer loop. A critical source of nonlinearity lies in the phase/frequency detection and the charge pump. The conversion of the phase error is not linearly translated to signals of other kinds and distortion is expected during the process. For a conventional CP-based synthesizer, the nonlinear charge pump and phase/frequency detector are the primary sources of nonlinearity encountered by the divider controller noise.

A typical tri-state PFD has the architecture shown in Fig. 2.9. The D flip-flops and the AND gate form the detector that outputs pulses that control the operation of two current sources having nominally identical currents. The current output enters the low-pass loop filter. Delays are usually used in the control circuit to avoid deadzone

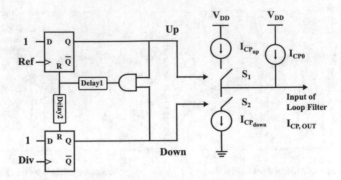

Fig. 2.9 The structure of a commonly used tri-state PFD and charge pump

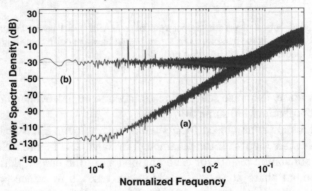

Fig. 2.10 Comparison between the accumulated quantization error which corresponds to the contributions to the output phase noise from the MASH 1-1-1 DDSM divider controller: (**a**) in the linear case and (**b**) in the presence of a PWL nonlinearity with 8% symmetrical mismatch

[24]. Due to mismatch between the up and down current sources, offset/leakage current, the delays that may exist in the detector, and a number of other possible nonlinearities [25], the input phase error is distorted when mapped to the input current to the loop filter. The overall PFD/CP nonlinearity is commonly modelled by a memoryless nonlinear transfer characteristic. In this work, as will be detailed, the nonlinearity of the PFD/CP is characterized by a piecewise linear model or a polynomial model.

The contribution to the output phase noise due to the divider controller becomes more significant because of nonlinear distortion. The accumulated quantization error, which corresponds to the phase deviation caused by the divider controller, will have higher low-frequency spectral content compared to the linear case; a flat noise floor can be observed. In addition, spurious tones appear in the spectrum and these tones will cause unwanted up/down conversion when the synthesizer is employed as a local oscillator. Figure 2.10 shows a comparison between the divider controller's accumulated quantization error $e_{acc}[n]$ and the distorted version of it, denoted $e_{acc}^{NL}[n]$, after experiencing a nonlinearity.

In this following parts of the section, two models of nonlinearities considered in this book, namely the PieceWise Linear (PWL) nonlinearity and the polynomial nonlinearity, are introduced. The characterization of the spur will be discussed and example simulation results for a nonlinear synthesizer will be presented.

2.3.2   Piecewise Linear Nonlinearity Model

The current output $I_{CP,OUT}$ from the PFD/CP structure shown in Fig. 2.9 can be modeled by a piecewise linear nonlinearity when considering only a simple mismatch between the current sources, i.e.,

$$I_{CP,OUT}[n] = b_0 + b_1 e_{acc}[n] + b_2 |e_{acc}[n] - e_{acc0}|, \tag{2.34}$$

where b_0 to b_2 are constants related to the tri-state PFD and CP parameters and e_{acc0} is the resulting offset in the Type-II synthesizer under consideration. In the feedforward model, most of the loop parameters only affect the shaping of the spectrum of the distorted accumulated quantization error. To generalize, the distortion is applied before the accumulated quantization error of the divider controller is shaped by the synthesizer loop, as shown in Fig. 2.7. Therefore, the distorted accumulated quantization error is found by [17, 18]

$$e_{acc}^{NL}[n] = \frac{N_{nom}}{I_{CP}} I_{CP,OUT}[n] \tag{2.35}$$

$$= b_0' + e_{acc}[n] + \frac{\epsilon}{2} |e_{acc}[n] - e_{acc0}|, \tag{2.36}$$

where

$$a_0' = \frac{1}{N_{nom}} \left[\left(\frac{T_{d1}}{T} \epsilon - \frac{T_{d2}}{T} \left(1 - \frac{\epsilon}{2}\right) \right) + \frac{I_{CP0}}{I_{CP}} \right], \tag{2.37}$$

$$e_{acc0} = \frac{a_0'}{1 - \frac{\epsilon}{2} \mathrm{sgn}(a_0')}, \tag{2.38}$$

$$b_0' = a_0' - e_{acc0}, \tag{2.39}$$

Here $\mathrm{sgn}(\cdot)$ is the signum function and T_{d1} and T_{d2} correspond to Delay1 and Delay2 in Fig. 2.9.

Empirically, in both measurements and simulations, when the nonlinearity has a point of discontinuity at zero, i.e., $a_0' = 0$, the worst spectral performance in terms of noise and spurious tones is observed. Therefore, in order to investigate the worst case scenario, a PWL nonlinearity in the form of

$$e_{acc}^{NL}[n] = e_{acc}[n] + \frac{\epsilon}{2} |e_{acc}[n]| \tag{2.40}$$

is considered.

The PWL nonlinearity is often used in simulations because it is a simple model which can exaggerate the possible nonlinear effects due to a discontinuity.

2.3.3 Polynomial Nonlinearity Model

Another type of nonlinearity model considered in this book is described by a polynomial expression [8], namely

$$e_{acc}^{NL}[n] = \sum_{k=0}^{p} c_k e_{acc}^{k}[n]. \tag{2.41}$$

The weights c_k can be determined by performing polynomial fitting to measured data. This means that the polynomial nonlinearity model can provide a more realistic description of a real-life synthesizer nonlinearity. However, the many degrees of freedom of polynomial fitting methods and the fitting strategy can lead to different weights in the polynomial. Examples of a PWL nonlinearity and a polynomial nonlinearity are shown in Fig. 2.11.

It is a common assumption that one or more of the simple polynomial terms are dominant and the performance of the divider controller is typically optimized based on such a simple polynomial term [8, 26]. An example connecting the PWL nonlinearity and polynomial nonlinearity from this perspective is detailed in [27], where the PWL nonlinearity with a symmetrical error in (2.40) can be approximated by

$$e_{acc}^{NL}[n] = e_{acc}[n] + \frac{\epsilon}{2} |e_{acc}[n]| \tag{2.42}$$

$$\approx e_{acc}[n] + \frac{\epsilon}{2 \max(e_{acc}[n])} e_{acc}^{2}[n]. \tag{2.43}$$

The dominant symmetrical error in the transfer characteristic of the PFD/CP is approximated by a second-order term. This approximation has been shown to be sufficiently accurate to estimate the divider controller-introduced noise floor [27].

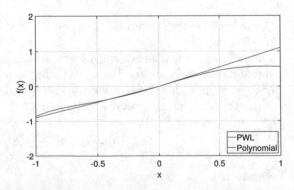

Fig. 2.11 Examples of PWL nonlinearity
$f(x) = x + 0.1|x|$ and polynomial nonlinearity
$f(x) = 1.03x + 0.12x^2 - 0.64x^3 - 0.27x^4 + 0.32x^5$ [8]

However, when the offset is non-zero in (2.36), the error part of the PWL nonlinearity is no longer symmetrical and the higher-order and odd-order terms in the polynomial approximation and their weights are different. A direct effect of this is an improvement in the divider controller phase noise contribution; the noise floor level and spurs both decrease in amplitude. This can be expected since the dominant term in the polynomial approximation changes with the offset.

Since $e_{acc}^{NL}[n]$ in the feedforward model determines the output phase noise contribution due to the divider controller, its performance can be regarded as the benchmark for the performance of the divider controller. This is based on the accuracy of the feedforward model with respect to the closed-loop method when estimating the divider controller-induced output phase noise in a conventional CP synthesizer [28, 29].

2.3.4 Measuring Spurious Tones

The nonlinearity has significant effects on the output phase noise contribution of the divider controller. One major spectral impact is nonlinearity-induced spurs. On the phase noise plot, the amplitude of a spur corresponds to the *energy* contained at the offset frequency. Since the periodic tone has constant power, the spur amplitudes increase with the length of the sequence used in the analysis. To compare spurious tones, it is conventional to obtain the power of them.

As introduced in the previous section, the modified periodogram and Welch's method are used to estimate the output phase noise or the spectra of phase-related quantities. These two methods are both biased estimators with identical expected values:

$$\mathbf{E}(P_x(Ff_s)) = \frac{1}{f_s NU} S_x(F) \star |W(F)|^2. \tag{2.44}$$

The asterisk denotes convolution. Consider a periodic signal

$$x[n] = A\cos(2\pi F_0 n), \tag{2.45}$$

where $F_0 \in [0, 0.5)$ and the expected value of the estimate is

$$\mathbf{E}(P_x(Ff_s)) = \frac{1}{f_s NU} \left(\frac{A^2}{4} \left(\sum_l \delta(F - F_0 + l) + \delta(F + F_0 + l) \right) \right) \star |W(F)|^2 \tag{2.46}$$

$$= \frac{1}{f_s NU} \frac{A^2}{4} \left(|W(F - F_0)|^2 + |W(F + F_0)|^2 \right). \tag{2.47}$$

The tone at $F_0 f_s$ is measured as

$$\mathbf{E}(P_x(F_0 f_s)) = \frac{1}{f_s N U} \frac{A^2}{4} |W(0)|^2. \tag{2.48}$$

The DC component $W(0)$ can be expressed by

$$W(0) = \sum_{n=0}^{N-1} w[n]. \tag{2.49}$$

With the definition of U in (2.33),

$$\mathbf{E}(P_x(F_0 f_s)) = \frac{A^2}{4} \frac{N}{f_s} \left(\frac{\frac{1}{N} \sum_{n=0}^{N-1} w[n]}{\sqrt{\frac{1}{N} \sum_{n=0}^{N-1} |w[n]|^2}} \right)^2 \tag{2.50}$$

Since the power in the peak of a double-sided power spectral density of a cosine wave is $A^2/4$, a factor of

$$G_P = \frac{f_s}{N} \left(\frac{\sqrt{\frac{1}{N} \sum_{n=0}^{N-1} |w[n]|^2}}{\frac{1}{N} \sum_{n=0}^{N-1} w[n]} \right)^2 = RBW \times \left(\frac{\text{rms}(w[n])}{\text{mean}(w[n])} \right)^2 \tag{2.51}$$

should be multiplied to the result of the power spectral density estimate for a spurious tone to obtain the power in the tone, which is expressed in the units of dBc.

2.3.5 Simulations with a Nonlinearity Model

With the synthesizer models and models for the nonlinearities, more realistic performance of a synthesizer can be simulated. The simulation results of the synthesizer with the parameters in Table 2.1 and with $\epsilon = 3\%$, $T_{d1} = 0.5$ ns, $T_{d2} = 0$ ns, and bleed current $I_{CP0} = 0$ are shown in Fig. 2.12. The reference path noise is -137 dBc/Hz and the VCO noise is -125 dBc/Hz. In Fig. 2.13, the divider controller phase noise contributions simulated with the closed-loop model and the feedforward model are shown.

The simulation results from the closed-loop behavioral model and the feedforward model are similar, both in terms of noise and spurs. The feedforward model

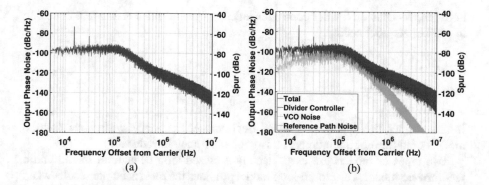

Fig. 2.12 Simulated output phase noise spectrum of a MASH 1-1-1-based fractional-N frequency synthesizer with the parameters in Table 2.1, a PWL nonlinearity of 3% mismatch, and identical noise contributors: (**a**) closed-loop behavioral model and (**b**) feedforward model

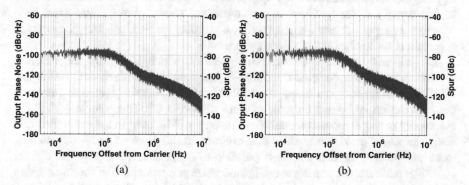

Fig. 2.13 Simulated divider controller phase noise contribution of a MASH 1-1-1-based fractional-N frequency synthesizer with the parameters in Table 2.1, a PWL nonlinearity of 3% mismatch, and identical noise contributors: (**a**) closed-loop behavioral model and (**b**) feedforward model. In-band spurs are measured to be -52.72 dBc, -62.82 dBc, and -70.19 dBc in (**a**) and -52.67 dBc, -64.23 dBc, and -69.76 dBc in (**b**)

can accurately predict the fundamental in-band spur and the noise floor level, with acceptable differences in the estimates of the harmonics of the fundamental spur in the presence of nonlinearities that are typically found in conventional CP synthesizers [28, 29]. The relatively accurate estimate of the output phase noise contribution from the divider controller with the feedforward model suggests that it is a valuable method for evaluating divider controller designs.

It should be cautioned again that the feedforward model is an *approximation* to the closed-loop model and care should be taken when interpreting predictions for strong nonlinearities.

As the results in Figs. 2.12 and 2.13 indicate, the spectral degradation due to the phase noise contribution of the divider controller can be significant. In this book, the focus is on divider controller-induced spectral phenomena that arise due to synthesizer nonlinearity.

2.4 Summary

In this chapter, the architecture of the fractional-N frequency synthesizer and the models used for the simulations in this book have been introduced.

Two types of models are used, namely a closed-loop behavioral model and a feedforward model. The closed-loop model provides the most accurate results when estimating divider controller-induced noise. The feedforward model accurately and efficiently determines the output phase noise contribution through the underlying mechanism of interaction between the accumulated quantization error and a weak memoryless nonlinearity. It is an approximate model that is valid for the degree of nonlinearity typically found in a conventional charge-pump synthesizer. Periodogram and periodogram-based methods are used to estimate the phase noise spectrum in the analysis of the simulation results.

Two models for the synthesizer nonlinearity, which is mainly the PFD/CP nonlinearity in a conventional fractional-N frequency synthesizer, are considered. The PWL nonlinearity is a simple model that is often used to characterize mismatch between the up and down currents of a tri-state PFD/CP. The other model is the commonly used polynomial model. Both models will be used in the evaluation and design of novel divider controller architectures in this book. The introduction of these nonlinearity models makes it possible to observe and compare the divider controller performance and the overall phase noise performance. For the comparison of the spur power, a window-dependent factor should be applied to the peridogram result.

The closed-loop model and the feedforward model yield similar results with identical simulation parameters in the cases presented. In general, the feedforward model is suitable for efficient evaluation of various divider controller designs for a charge-pump fractional-N frequency synthesizer. Since the feedforward model is an approximate model, care must be taken when extrapolating to cases where stronger nonlinearities are present.

Chapter 3
Spurious Tones in Fractional-N Frequency Synthesizers

As outlined in Chap. 2, the output phase noise contribution of the divider controller in the nonlinear case is a cause of the spurious tones in a fractional-N frequency synthesizer. In this chapter, a brief classification of the observed stationary spurs related to the divider controller noise is first presented. Since time-varying phenomena may not be observed using the long-term spectrum, wandering spurs are usually analyzed using the spectrogram method. An introduction to this method follows as the second part of the chapter.

3.1 Divider Controller-Induced Stationary Spurs in a MASH-Based Frequency Synthesizer

The commonly used MASH DDSM divider controller architecture has been found to be responsible for many types of spurious tones in the output phase noise of a fractional-N frequency synthesizer. In this section, three types of fixed spurs that exist in a MASH-based fractional-N frequency synthesizers are discussed: integer boundary spurs, fractional boundary spurs, and sub-harmonic spurs.

3.1.1 Integer Boundary Spurs (IBS)

Integer boundary spurs appear when the input to the divider controller is close to 0 or the modulus M. When a small input X is applied to the MASH-based divider controller, a primary IBS can be expected at

$$f_{IBS} = \frac{X}{M} f_{PFD}.$$

(3.1)

The cause of the spur is the underlying pattern hidden in the waveform of the accumulated quantization error $e_{acc}[n]$ of the divider controller. A sawtooth pattern

$$T_{frac}[n] = \frac{1}{M} \left(M - X(n - \delta)\right) \bmod M, \tag{3.2}$$

where δ is an offset, can be identified in the waveform of $e_{acc}[n]$ [30].

An example of the waveform and the spectrum of $e_{acc}[n]$ and $e_{acc}^{NL}[n]$ of a third-order MASH DDSM in the presence of a PWL nonlinearity with $X = 3000$, $M = 2^{24}$, and $f_{PFD} = 122.88$ MHz are shown in Fig. 3.1. Despite the similarity of the time domain waveforms, $e_{acc}[n]$ has a characteristic second-order high-pass shaped spectrum while $e_{acc}^{NL}[n]$ exhibits an elevated noise floor and fixed spurs. The expected fundamental spur (the IBS), according to (3.1), is at 21.97 kHz.

When the input is close to the modulus M of the divider controller, the input can be expressed as

$$X = M - X^*. \tag{3.3}$$

A pattern of

$$T_{frac}[n] = \frac{1}{M} \left(X^*(n - \delta)\right) \bmod M \tag{3.4}$$

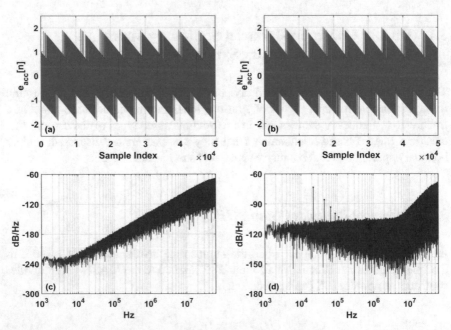

Fig. 3.1 The waveforms and the spectra of (**a**), (**c**) $e_{acc}[n]$ and (**b**), (**d**) $e_{acc}^{NL}[n]$ in the presence of an 8% PWL nonlinearity for a MASH 1-1-1 divider controller with input $X = 3000$, $M = 2^{24}$, and $f_{PFD} = 122.88$ MHz. The sawtooth $T_{frac}[n]$ is superimposed on the $e_{acc}[n]$ waveform in red

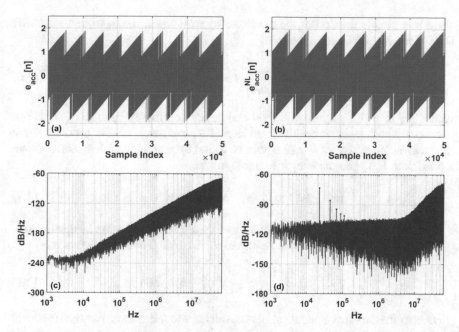

Fig. 3.2 The waveforms and the spectra of (**a**), (**c**) $e_{acc}[n]$ and (**b**), (**d**) $e_{acc}^{NL}[n]$ in the presence of an 8% PWL nonlinearity for a MASH 1-1-1 divider controller with input $X = M-3000 = 16774216$, $M = 2^{24}$, and $f_{PFD} = 122.88$ MHz. The sawtooth $T_{frac}[n]$ is superimposed on the $e_{acc}[n]$ waveform in red

can be recognized in the $e_{acc}[n]$ waveform in this case. The nonlinearity emphasizes an underlying pattern at this frequency, giving rise to the spurious tones. The fundamental offset frequency (from the carrier) of the IBS in this case is

$$f_{IBS} = \frac{X^*}{M} f_{PFD}. \tag{3.5}$$

A example with input $X = M - 3000$ is shown in Fig. 3.2.

In both cases, harmonics of the fundamental IBS appear in the spectrum of $e_{acc}^{NL}[n]$; consequently, as explained in Chap. 2, they also appear in the output phase noise contribution of the divider controller.

3.1.2   Fractional Boundary Spurs

The frequency of the fundamental IBS is predicted to be outside the bandwidth of the fractional-N frequency synthesizer when the input is sufficiently far away from 0 and the modulus M of the divider controller. However, significant in-band spurs could still exist. In-band spurs can appear if the input of the divider controller is close to a simple fraction of the modulus. In such cases, the input to the MASH DDSM can be expressed in the form

$$X = \frac{X_{Num}}{X_{Den}} M = \frac{aM}{b} + X', \tag{3.6}$$

where X_{Num} and X_{Den}, as well as a and b, are coprime integers [31]. X' is not necessarily an integer or positive, and should have a much smaller absolute value than aM/b. The value of b in (3.6) can be found by searching all integers less than a maximum b_{max} and the integer b should minimize

$$\min \{(bX) \bmod M, \ (M - bX) \bmod M\}. \tag{3.7}$$

The found minimum of (3.7) is $b|X'|$ and therefore X' and a can be determined. The pattern in $e_{acc}[n]$ that contributes to the fractional boundary spurs is

$$T_{frac}[n] = \frac{1}{M} \left(M - X'(n - \delta) \right) \bmod M. \tag{3.8}$$

Due to the fact that b identical patterns exist, the fundamental of the fractional boundary spurs is at

$$f_{FBS} = \frac{bX'}{M} f_{PFD}. \tag{3.9}$$

In the literature, fractional boundary spurs are also called 'fractional spurs' [31]. By formatting X' as

$$X' = \frac{cM}{D}, \tag{3.10}$$

the frequency of the fundamental fractional boundary spur can be expressed as

$$f_{FBS} = \frac{bc}{D} f_{PFD}. \tag{3.11}$$

Due to the fact that the least common multiple $LCM(b, D) = D$, the fractional boundary spurs can be categorized as 'fractional spurs' at offset frequencies of [31]

$$f_k = \frac{k}{X_{Den}} f_{PFD}, \ k = 1, \ 2, \ \dots. \tag{3.12}$$

An example with $X = M/2 + 1024$, $M = 2^{24}$, and $f_{PFD} = 122.88$ MHz is shown in Fig. 3.3. The fundamental of the fractional boundary spur is at a 15 kHz offset from the carrier, as predicted by (3.9).

Fig. 3.3 (**a**) The waveform of $e_{acc}[n]$, (**b**) the waveform of $e_{acc}^{NL}[n]$, and (**c**), (**d**) the spectra $e_{acc}^{NL}[n]$ in the presence of an 8% PWL nonlinearity for a MASH 1-1-1 with input $X = M/2 + 1024 = 8389632$, $M = 2^{24}$, and $f_{PFD} = 122.88$ MHz. The sawtooth $T_{frac}[n]$ is superimposed on the $e_{acc}[n]$ waveform in red. The red dots in (**c**) and (**d**) indicate the fractional boundary spurs and its harmonics. The initial condition $s_1[0] = 1$ in (**a**), (**b**), and (**c**) and $s_1[0] = 171$ in (**d**)

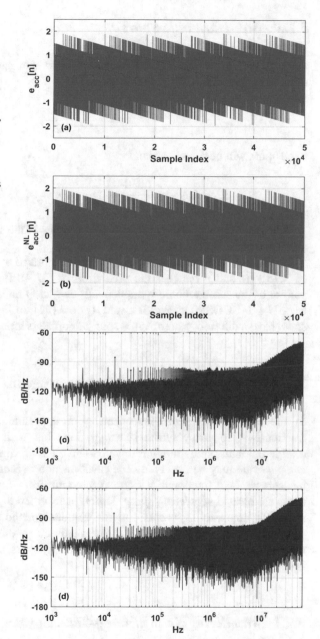

3.1.3 Sub-Harmonic Spurs

Another type of MASH DDSM-induced in-band spur can be observed. When the input is in the form (3.11) with $b = 2^x$ and $D = 2^y$:

$$X = \frac{aM}{2^x} + \frac{cM}{2^y},$$ (3.13)

fixed spurs will be present around

$$f_{sub,k} = \frac{(2k+1)}{2D} f_{PFD}, \; k = 0, \; 1, \; 2, \; \ldots$$ (3.14)

Since these spurs appear at fractional multiples of the spur offsets in (3.12), they are also termed 'sub-fractional' spurs in [31].

In Fig. 3.3, a pair of sub-harmonic spurs is centered at a fundamental frequency of 3.75 kHz and it matches the prediction of (3.14). It should be noted that the pattern of the sub-harmonic spurs depends explicitly on the initial condition of the MASH 1-1-1 DDSM, as Fig. 3.3c and d show [32]. In Fig. 3.3d, the sub-harmonic spurs have a different pattern and a spur pair appears around 3.75 kHz.

3.2 Wandering Spurs and the Spectrogram Method

In the output phase noise spectrum of a conventional MASH DDSM-based fractional-N frequency synthesizer, especially one with a MASH DDSM having a large modulus, spurs moving towards and away from the carrier frequency at a *constant* rate may be observed. The phenomenon is seen both in simulations and measurements of fractional-N frequency synthesizers. Such time-varying spurious tones are termed *wandering spurs*. Different from fixed-position spurs, wandering spurs are time-varying events and the standard method of spectral analysis for stationary spurs is not suitable to observe them. In this section, an observation method that facilitates the study of wandering spurs, namely the spectrogram, is described.

3.2.1 Wandering Spur in Long-Term Spectrum

Traditionally, a long data sequence is used to identify spurs at stationary offset frequencies from the carrier. In Fourier-based methods, the spur amplitudes are generally proportional to the length of the data sequence used in the analysis. The long-term spectrum is therefore an appropriate method to evaluate the performance of a synthesizer in terms of spurs.

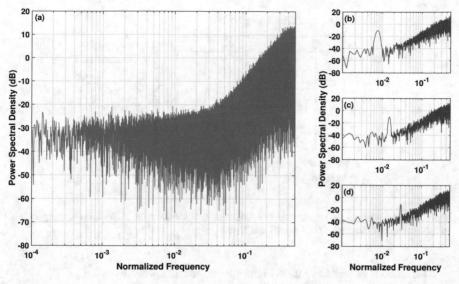

Fig. 3.4 (**a**) A simulated long-term spectrum (calculated from 5×10^5 points) of a wandering spur event and (**b**), (**c**), (**d**) short-term spectra (calculated from 4000-point segments) of the same data with different offsets from the beginning from the sequence

However, the long-term spectrum is not informative when short-term events take place in the spectrum of interest. In the case of time-varying spectral events, this method simply shows the *average* effect of the events within the observation window. This is especially true in the case of wandering spurs. Since the peaks of the spurs move across the phase noise spectrum, the long-term spectrum will implicitly perform averaging and interpret the events as noise, as shown in Fig. 3.4a.

In order to study the wandering spur phenomenon, numerous short-term 'snapshots' (Fig. 3.4b–d) are needed. This is inconvenient in further analysis of the pattern of the spurs, e.g., identifying the moving rates of the spurs and the period of events.

3.2.2 Spectrogram Method

Therefore, a method to display time-varying phenomena in the short-term spectrum is required. In this book, the spectrogram, i.e., successive snapshots of the spectrum, is used to analyze the evolution of output phase noise and related quantities. This technique is implemented in commercial spectrum analyzers with a real-time measurement mode [33].

Assume that a data sequence $x[n]$ contains N samples. Denote the window length of the short-term spectral analysis as N_w and, for better use of the data sequence, an overlap of D is considered when shifting the window over the data sequence, as illustrated in Fig. 3.5.

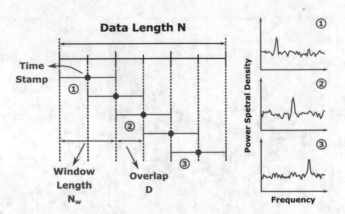

Fig. 3.5 Illustration of the spectrogram method used to observe events in the short-term spectrum. In this example, $N = 3N_w$, $D = N_w/2$, and $L = 5$

The number of windows used in the spectrogram is given by

$$L = \left\lfloor \frac{N - D}{N_w - D} \right\rfloor \tag{3.15}$$

where $\lfloor \cdot \rfloor$ is the floor function.

The time stamp to which each window or spectrum corresponds is defined by the time at the middle of the window. The time stamp of the kth window is thus

$$\frac{1}{f_s} \left(\frac{N_w}{2} + (k - 1)N_w \right), \tag{3.16}$$

where f_s is the sampling frequency of the data sequence. When the analyzed sequence has no specific sampling frequency, $f_s = 1$ will be assumed and the time stamp values correspond to the sample indices.

For each analysis window, conventional spectral analysis techniques can be applied. In this book, a modified periodogram is applied to each window [22, 23].

To express the result that contains three quantities, namely time, frequency, and amplitude, in a two-dimensional figure, a color map is used. In this book, red and colors closer to red represent higher amplitudes and blue and colors closer to blue represent lower amplitudes. An example spectrogram is shown in Fig. 3.6. The typical vee-shaped pattern of a wandering spur event is visible. The subfigures on the right are individual short-term spectra at three marked time instants, i.e., they are the two-dimensional slices of the spectrogram.

Fixed spurs can also be observed in the spectrogram. As shown in Fig. 3.7, the spur at a fixed frequency appears as a vertical line on the spectrogram. In the example shown, a spur at a normalized frequency of approximately 0.025 is present and is shown as a red line.

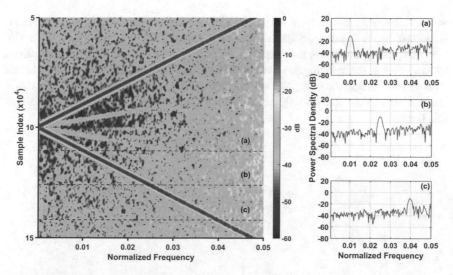

Fig. 3.6 An example simulated spectrogram with short-term spectra at three time instants (**a**), (**b**), and (**c**) shown. The spectrogram consists of spectra computed from segments of 4000 points

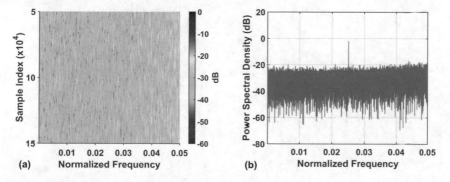

Fig. 3.7 An example of (**a**) a simulated spectrogram with a fixed spur and (**b**) the long-term spectrum showing this spur

3.3 Summary

Spurs are always a concern when characterizing the performance of a fractional-N frequency synthesizer. Three types of nonlinearity-induced fixed spurs within a MASH-based fractional-N frequency synthesizer can be identified: integer boundary spurs, fractional boundary spurs/fractional spurs, and sub-harmonic spurs. The nonlinearity extracts the frequency components corresponding to the underlying patterns that lead to the most common spurious tones seen in fractional-N frequency synthesizers.

Wandering spurs are time-varying spurious tones that are observed in the short-term spectrum of a fractional-N frequency synthesizer. Unlike stationary spurs, wandering spurs move in frequency. The long-term spectrum that is used to evaluate stationary spurs is not suitable for the observation of wandering spurs since the averaging effect interprets these spurs as noise. To observe and analyze such time-varying events, the spectrogram method is used. The spectral analysis window shifts with an overlap and the analysis result is displayed in a two-dimensional plot with a color scheme. This method will be used in the study of wandering spurs throughout this book.

Chapter 4
Wandering Spurs and the MASH DDSM Divider Controller

In this chapter, the root cause of the wandering spur phenomenon is elaborated. The analysis starts with simulations that reproduce experimentally observed wandering spur events. These events can be traced back to the commonly-used MASH 1-1-1 divider controller.

4.1 MASH 1-1-1 DDSM-Based Fractional-N Frequency Synthesizer

Wandering spurs can be reproduced in idealized simulations of a fractional-N frequency synthesizer. As mentioned in Chap. 2, the MASH DDSM is widely used as the divider controller because of its high-pass shaped quantization noise. The third-order MASH 1-1-1 is often used due to its sufficiently randomized quantization error. The structure of a MASH 1-1-1 is shown in Fig. 4.1. In a local oscillator application the input is kept constant, i.e., $x[n] = X$. A first-order LSB dither $d[n]$ is applied to randomize the quantization noise and it ensures that the output of the MASH is spur-free.

The quantization error and its delayed version in the ith stage are denoted as $e_i[n]$ and $s_i[n]$, respectively. The outputs of the three first-order error feedback modulators (EFM1s) are related to the quantization errors by

$$y_1[n] = \frac{1}{M} \left(X - (e_1[n] - e_1[n-1]) \right), \tag{4.1}$$

$$y_2[n] = \frac{1}{M} \left(e_1[n] - (e_2[n] - e_2[n-1]) + d[n] \right), \tag{4.2}$$

$$y_3[n] = \frac{1}{M} \left(e_2[n] - (e_3[n] - e_3[n-1]) \right). \tag{4.3}$$

© The Author(s), under exclusive license to Springer Nature Switzerland AG 2022
D. Mai, M. P. Kennedy, *Wandering Spurs in MASH-based Fractional-N Frequency Synthesizers*, Analog Circuits and Signal Processing,
https://doi.org/10.1007/978-3-030-91285-7_4

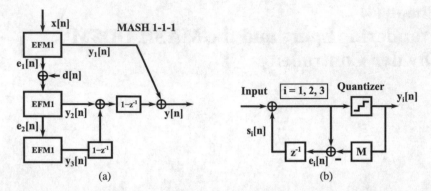

Fig. 4.1 Block diagrams of (**a**) a MASH 1-1-1 and (**b**) a first-order error feedback modulator (EFM1). For a conventional LSB-dithered MASH 1-1-1, a 1-bit dither $d[n]$ is introduced to reduce the periodicity in the quantization error

The internal states of the EFM1s can be expressed by

$$e_1[n] = s_1[n+1] = (s_1[0] + (n+1)X) \mod M, \tag{4.4}$$

$$e_2[n] = s_2[n+1] = \left(s_2[0] + \sum_{m=0}^{n} (e_1[m] + d[m]) \right) \mod M, \tag{4.5}$$

$$e_3[n] = s_3[n+1] = \left(s_3[0] + \sum_{m=0}^{n} e_2[m] \right) \mod M, \tag{4.6}$$

where $s_i[0]$ is the initial condition of the ith stage.

4.2   Observation of Wandering Spurs in a Fractional-N Synthesizer

Wandering spurs have been observed in a measured spectrogram of a commercial fractional-N frequency synthesizer [34]. An example measured spectrogram of the output phase noise of a commercial fractional-N frequency synthesizer is shown in Fig. 4.2. The measurement was taken from a ADF4159EB1Z evaluation board, which contains a 100 MHz reference crystal oscillator, an ADF4159 frequency synthesizer, and a HMC515 VCO. The ADF4159 frequency synthesizer contains a third-order MASH DDSM divider controller [35, 36].

As the example measurement shows, during a typical wandering spur event, the spurs first move with a constant rate towards the carrier and then subsequently away from it. This results in the characteristic X-shaped pattern in the spectrogram of the

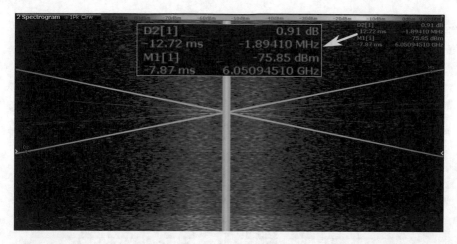

Fig. 4.2 Measured spectrogram of the output phase noise of ADF4159 frequency synthesizer on ADF4159EB1Z evaluation board. The spectrogram shows a typical wandering spurs event around the carrier frequency

output phase noise. The example wandering spur event shown in Fig. 4.2 repeats with a constant period during the observation.

To investigate the cause of wandering spurs observed in measurements, the phenomenon is reproduced qualitatively in simulations. Simulation results for a MASH 1-1-1-based synthesizer with the parameters in Table 2.1 and an input $X = 8$ are shown in Fig. 4.3. A PWL nonlinearity modeling a static mismatch of 8% between the up and down current sources in the charge pump is applied.

Note that the timing of the wandering spur events matches the repeating pattern in the waveform of the VCO input voltage. The timing of the events also corresponds to those when the hidden sawtooth waveform in the accumulated quantization error $e_{acc}[n]$ changes abruptly in value. Since this hidden pattern is related to the accumulation of the constant input, the wandering spur events also correspond to the abrupt changes in the sawtooth waveform of $e_1[n]$.

Zoomed-in views of a single event in Fig. 4.3 are shown in Fig. 4.4. The chirping behavior of the control voltage can be observed at the input to the VCO, which is shown in Fig. 4.4b: the waveform pattern first decreases in frequency and becomes almost steady in amplitude; then its frequency starts to increase. This leads to the spur moving first towards the zero offset frequency and then away from it in the phase noise contribution of the MASH 1-1-1 shown in Fig. 4.4a. Since the output phase noise contribution due to the MASH 1-1-1 is determined by the accumulated quantization error, a similar chirping pattern in the waveform of $e_{acc}[n]$ can be observed; this is shown in Fig. 4.4c. Examining all EFM1 stage outputs within the MASH 1-1-1, it should be noticed that the second stage output $y_2[n]$ exhibits similar chirping behavior, as shown in Fig. 4.4d.

The results of the simulation suggest that wandering spurs originate in the MASH divider controller. In many cases, the wandering spurs in the spectrogram exhibit a

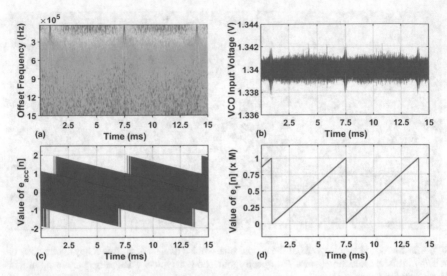

Fig. 4.3 Simulated (**a**) spectrogram of the output phase noise contribution of the MASH 1-1-1 showing the wandering spur events, (**b**) the waveform of the VCO input voltage, (**c**) the waveform of the accumulated quantization error $e_{acc}[n]$, and (**d**) the internal state $e_1[n]$ of the MASH divider controller. Input is $X = 8$ and the modulus of the MASH 1-1-1 is $M = 2^{20}$

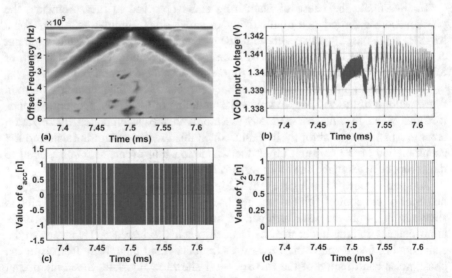

Fig. 4.4 Simulated (**a**) output phase noise contribution of the MASH 1-1-1 showing the wandering spur events, (**b**) the waveform of the VCO input voltage, (**c**) the waveform of the accumulated quantization error $e_{acc}[n]$ from the MASH divider controller, and (**d**) second-stage output $y_2[n]$ waveform around a single wandering spur event. Input is $X = 8$ and the modulus of the MASH is $M = 2^{20}$

constant rate of linear movement, as shown in Figs. 4.3a and 4.4a. This constant moving rate is termed *wander rate*. Typically, the wandering spur events also repeat with a constant period, as shown in Fig. 4.3a.

4.3   Generating Mechanism of Wandering Spur in a MASH 1-1-1-Based Fractional-N Synthesizer

In the example shown in the previous section, the output of the MASH 1-1-1 second stage EFM1 $y_2[n]$ has chirping behaviors that coincide with the wandering spur events. The spectrogram of the signal $y_2[n]$ can be plotted and an example with a small MASH 1-1-1 input $X = 3$ is shown in Fig. 4.5a. The spectrogram shows a characteristic pattern of wandering spurs with different wander rates.

The expression for $y_2[n]$ can be written as

$$y_2[n] \approx \frac{1}{M}\left(e_1[n] - (e_2[n] - e_2[n-1])\right) \tag{4.7}$$

in the time domain and

$$Y_2(z) \approx \frac{1}{M}\left(E_1(z) - \left(1 - z^{-1}\right)E_2(z)\right) \tag{4.8}$$

in the frequency domain. The approximation neglects the effect of the 1-bit dither $d[n]$. Since the EFM1 is a digital accumulator, in the first EFM1

$$e_1[n] = s_1[n+1] = (s_1[0] + (n+1)X) \mod M. \tag{4.9}$$

This means that the signal $e_1[n]$ has a simple sawtooth waveform. It should manifest itself in the frequency domain as stationary tones. Therefore, the high-pass

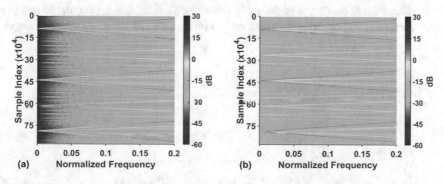

Fig. 4.5 Spectrogram of (**a**) $y_2[n]$ and (**b**) $e_2[n]$, input is $X = 3$ and the MASH 1-1-1 modulus is $M = 2^{20}$

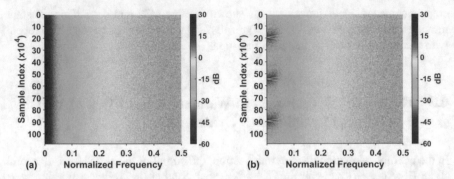

Fig. 4.6 Simulated spectrogram of (**a**) $e_{acc}[n]$ and (**b**) $e_{acc}^{NL}[n] = e_{acc}[n] + 0.03e_{acc}^2[n]$. Input is $X = 3$ and the modulus of the MASH is $M = 2^{20}$. Note that $e_{acc}[n]$ does not exhibit wandering spurs

filtered $(1 - z^{-1})E_2$, or essentially $e_2[n]$, contributes to the pattern seen in the $y_2[n]$ spectrogram. The simulated spectrograms of $y_2[n]$ and $e_2[n]$ in a MASH 1-1-1 for $X = 3$ case are shown in Fig. 4.5. In this case, the fixed tones with a fundamental frequency of $3/M$ are too close to zero normalized frequency to be observed in the spectrogram.

The contribution of the phase noise from the MASH 1-1-1 is directly related to the accumulated quantization error, as the feedforward model in Chap. 2 suggests. The simulation results for the $e_{acc}[n]$ waveform in the linear case and in the presence of a nonlinearity are shown in Fig. 4.6. These results indicate that, in this example case, there are no wandering spurs in $e_{acc}[n]$ itself; they arise in $e_{acc}^{NL}[n]$ due to the nonlinearity.

4.3.1   Linear Case

In the linear case, namely a completely linear synthesizer, the accumulated quantization error $e_{acc}[n]$ contributes to the output phase noise directly. The expression of $e_{acc}[n]$ can be written as

$$
e_{acc}[n] = \sum_{k=0}^{n-1} \left(y[k] - \frac{X}{M} \right)
$$

$$
= \sum_{k=0}^{n-1} y_1[k] + y_2[n-1] + \nabla y_3[n-1] - \frac{nX}{M}
$$

$$
= -\frac{1}{M} \left(\nabla^2 e_3[n-1] - s_1[0] - d[n-1] \right), \tag{4.10}
$$

where ∇ and ∇^2 denote the first and second-order backward difference operators, i.e.,

$$\nabla y_3[n-1] = y_3[n-1] - y_3[n-2], \tag{4.11}$$

$$\nabla^2 e_3[n] = \nabla e_3[n] - \nabla e_3[n-1]. \tag{4.12}$$

Since the initial condition $s_1[0]$ contributes to the DC value of $e_{acc}[n]$, which will be overcome by the Type-II loop considered and consequently does not affect the phenomenon across the spectrum, this term can be neglected. The 1-bit LSB dither, which is of negligible amplitude compared to $\nabla^2 e_3[n]$, can also be ignored in the analysis due to its limited spectral impact. Therefore,

$$e_{acc}[n] \approx -\frac{1}{M}\nabla^2 e_3[n-1]. \tag{4.13}$$

Based on the operation of the EFM1, ∇e_3 is associated with $e_2[n]$ by

$$\nabla e_3[n] = \begin{cases} e_2[n], & \text{if } e_2[n] + e_3[n-1] < M, \\ e_2[n] - M, & \text{if } e_2[n] + e_3[n-1] \geq M. \end{cases} \tag{4.14}$$

Therefore, the value of $e_{acc}[n]$, which is mainly determined by $\nabla^2 e_3[n]$, can be related to

$$\nabla e_2[n] = e_2[n] - e_2[n-1] \tag{4.15}$$

based on the conditions of $e_2[n]$ and $e_3[n]$. From this perspective, $e_{acc}[n]$ can take on one of three possible values

$$e_{acc}[n] \approx \begin{cases} -\frac{1}{M}\nabla e_2[n-1], \\ -\frac{1}{M}\nabla e_2[n-1] - 1, \\ -\frac{1}{M}\nabla e_2[n-1] + 1. \end{cases} \tag{4.16}$$

$$= -\frac{1}{M}\nabla e_2[n-1] + \nabla y_3[n-1]. \tag{4.17}$$

From the definition of $y_3[n]$,

$$\nabla y_3[n] = \frac{1}{M}\left(\nabla e_2[n] - \nabla^2 e_3[n]\right). \tag{4.18}$$

Substituting (4.18) into (4.17) yields (4.13). In particular, the $e_2[n]$-related term is canceled in $e_{acc}[n]$. The shaped third stage quantization error is the only dominant spectral component. No wandering spur pattern is observed in the linear case, as

Fig. 4.6a shows. The third stage quantization error $e_3[n]$ can be approximated by uniformly distributed white noise in this case.

4.3.2   Nonlinear Case

Note, however, that the wandering spur patterns appear when nonlinearity is present, as shown in Fig. 4.6b. A polynomial fit can be performed to characterize real-life nonlinearities and therefore the polynomial may be considered as a general model [8, 37]. The distorted accumulated quantization error is a sum of terms containing $e_{acc}[n]$ raised to different powers.

Consider a typical term $e_{acc}^p[n]$ in a polynomial nonlinear distortion. The second-order difference of $e_3[n]$, which is approximately the accumulated quantization error $e_{acc}[n]$ following (4.13), shows a correlation with $e_2[n]$ when experiencing an even-order distortion. For example, in the input $X = 3$ case, the normalized cross-correlations between $\nabla^2 e_3[n]$ and $e_2[n]$ and that between $(\nabla^2 e_3)^2[n]$ and $e_2[n]$ are shown in Fig. 4.7. The correlation between $\nabla^2 e_3[n]$ and $e_2[n]$ is negligible but the nonlinearly distorted version $(\nabla^2 e_3)^2[n]$ is correlated with $e_2[n]$.

When the second-order difference of $e_3[n]$ experiences an odd-order distortion, it shows greater correlation with $\nabla e_2[n]$ than with $e_2[n]$. An example with the input $X = 3$ is shown in Fig. 4.8. Again, $\nabla^2 e_3[n]$ has negligible correlation with $e_2[n]$. The distorted version $(\nabla^2 e_3)^3[n]$ shows greater correlation with $\nabla e_2[n]$. The discrepancy between the even and odd-order cases can be explained from the perspective of their effects on the time-domain waveforms. The even-order distortion causes the negative terms in $\nabla^2 e_3[n]$ to become positive after distortion, leading to a similar structure to $e_2[n]$, which is locally line-symmetric around a wandering spur event (as shown in Fig. 4.9a and c). The odd-order distortion preserves the polarity of the terms of $\nabla^2 e_3[n]$ and therefore the distorted $\nabla^2 e_3[n]$

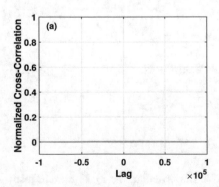
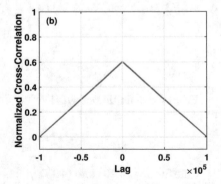

Fig. 4.7 Simulated normalized cross-correlation between (**a**) $\nabla^2 e_3[n]$ and $e_2[n]$ and (**b**) $(\nabla^2 e_3)^2[n]$ and $e_2[n]$. Input is $X = 3$ and the modulus of the MASH is $M = 2^{20}$. Sequences with identical lengths at the same offset from the beginning of the data are used

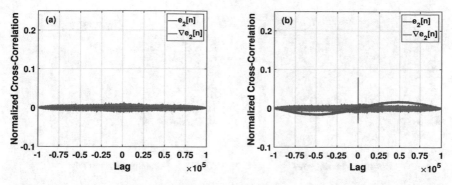

Fig. 4.8 Simulated (**a**) normalized cross-correlation between $\nabla^2 e_3[n]$ and $e_2[n]$ and normalized cross-correlation between $\nabla^2 e_3[n]$ and $\nabla e_2[n]$ and (**b**) normalized cross-correlation between $(\nabla^2 e_3)^3[n]$ and $e_2[n]$ and normalized cross-correlation between $(\nabla^2 e_3)^3[n]$ and $\nabla e_2[n]$. Input is $X = 3$ and the modulus of the MASH is $M = 2^{20}$. Sequences with identical lengths at the same offset from the beginning of the data are used

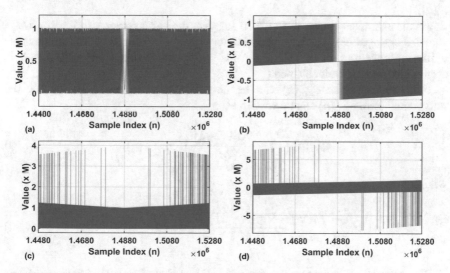

Fig. 4.9 Simulated waveforms of (**a**) $e_2[n]$, (**b**) $\nabla e_2[n]$, (**c**) $(\nabla^2 e_3)^2[n]$, and (**d**) $(\nabla^2 e_3)^3[n]$ showing local symmetry. Input is $X = 3$ and the modulus of the MASH is $M = 2^{20}$

exhibits approximate point symmetry locally, which resembles the $\nabla e_2[n]$ sequence in part (as Fig. 4.9b and d show),

Since the wandering spur pattern in $\nabla e_2[n]$ has low amplitudes at low frequencies, the dominant low-frequency wandering spurs are associated with even-order distortions. For a typical nonlinearity within a synthesizer, a prominent linear contribution of $e_{acc}[n]$ from the MASH 1-1-1, which is second-order shaped, is present. Considering the limited contribution of the odd-order terms in a typical nonlinearity compared to the linear contribution, the approximately first-order shaped

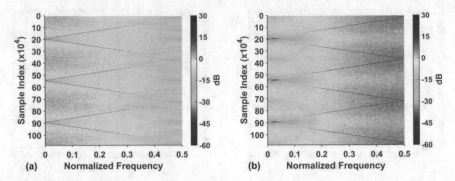

Fig. 4.10 Simulated spectrogram of (**a**) $(\nabla^2 e_3[n]/M)^2$ and (**b**) $(\nabla^2 e_3[n]/M)^3$. Input is $X = 3$ and the modulus of the MASH is $M = 2^{20}$

spur pattern is generally not significant at high frequencies in the spectrogram of $e_{acc}^{NL}[n]$.

As the examples shown in Fig. 4.10 suggest, the spectrogram of $(\nabla^2 e_3[n]/M)^2$ shows a spur pattern that has approximately equal amplitudes at low frequencies while the spectrogram of $(\nabla^2 e_3[n]/M)^3$ has a wandering spur pattern that is approximately first-order shaped.

Furthermore, since $e_{acc}[n]$ can be approximately expressed by (4.17), the polynomial term $e_{acc}^p[n]$ can be written as

$$e_{acc}^p[n] \approx \left(-\frac{\nabla e_2[n-1]}{M}\right)^p + T_{e_2}[n], \tag{4.19}$$

where

$$T_{e_2}[n] = \sum_{m=0}^{p-1} \binom{p}{m} \left(-\frac{\nabla e_2[n-1]}{M}\right)^m (\nabla y_3[n-1])^{p-m}. \tag{4.20}$$

The correlation between the distorted $\nabla^2 e_3[n]$ and $e_2[n]$-related sequences indicates that the summation in (4.19) does not produce a signal that is uncorrelated with $e_2[n]$ or $\nabla e_2[n]$. The spectral pattern in $e_2[n]$ or $\nabla e_2[n]$ will therefore be manifest in part in the spectrogram of the distorted accumulated quantization error $e_{acc}^{NL}[n]$.

The underlying pattern in the $e_2[n]$ spectrogram has wandering spur events at different wander rates,[1] as shown in the example in Fig. 4.5b. The wandering spurs in $e_2[n]$ contain components at a range of frequencies. Assume that the energy in the signal in a window is constant and that the $e_2[n]$ waveform causes similar wandering

---

[1] An event in the $e_2[n]$ spectrogram can contain harmonics. The wander rate here refers to the moving rate of the wandering spur at the fundamental frequency.

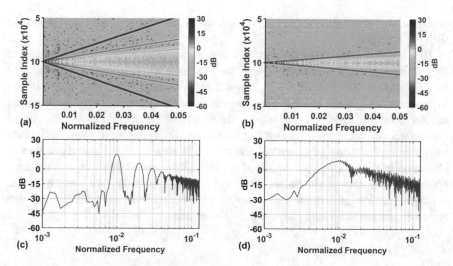

Fig. 4.11 Simulated spectrogram and example spectra of $e_2[n]/M$ when the MASH 1-1-1 input is (**a**), (**c**) $X = 1$ and (**b**), (**d**) $X = 4$. The modulus of the MASH is $M = 2^{20}$

spur harmonic patterns, then the wandering spurs in an event with a higher wander rate will have wider spreads of frequency components and thus the spur amplitudes will be lower. Examples of simulated most prominent wandering spur events in their $e_2[n]/M$ spectrograms are shown in Fig. 4.11. The event with a higher wander rate (right) consists of spurs with lower amplitudes. This conclusion also applies to a single $e_2[n]$ spectrogram, for example in Fig. 4.5b. In the $e_2[n]$ spectrogram, events with lower wander rates have higher wandering spur amplitudes. A similar conclusion applies to the pattern in the $\nabla e_2[n]$ spectrogram.

It should be noted that $e_2[n]$ or $\nabla e_2[n]$ does not directly contribute to $e_{acc}^p[n]$. Rather, a component that can be associated with $e_2[n]$ or $\nabla e_2[n]$ in the result of the summation in (4.19) leads to the wandering spur pattern in the presence of nonlinearity. Noise that is free of wandering spur patterns exists and only the most prominent wandering spur in the underlying pattern, i.e., the wandering spurs at the slowest wander rate, will typically be observed in the output phase noise spectrogram.

4.4 Summary

The wandering spur phenomenon originates in the contribution of the MASH DDSM divider controller to the output phase noise. Tracing backwards from the VCO output phase noise, the chirping behavior that corresponds to the wandering spur events can be observed in the VCO input voltage, the accumulated quantization error of the MASH 1-1-1 divider controller, and the second stage output $y_2[n]$ within the MASH. Since the first stage input does not contribute to the frequency-varying component, the internal stage $e_2[n]$ causes the wandering spur pattern seen in the $y_2[n]$ spectrogram.

In a general case where the second stage EFM quantization error exhibits the underlying pattern of wandering spurs—for example, the small input case presented in this chapter—the MASH 1-1-1 does not contribute to wandering spur patterns in the case of a *linear* synthesizer. The cancellation between stages leads to the accumulated quantization error, which contributes to the output phase noise, being dominated by a clean and shaped quantization error related to the third stage quantization error $e_3[n]$. In the nonlinear case, however, an $e_2[n]$-related component shows up in the distorted accumulated quantization error, giving rise to wandering spurs in the output phase noise. Since $e_2[n]$ does not contribute directly to the distorted accumulated quantization error, only the slowest wandering spurs are typically seen in the output phase noise spectrogram.

Chapter 5
Analysis of MASH 1-1-1-Induced Wandering Spur Patterns in a Fractional-N Frequency Synthesizer

In this chapter, the analysis of the wandering spur patterns of a MASH 1-1-1-based fractional-N frequency synthesizer is presented. The internal states of a MASH 1-1-1 that cause the wandering spur patterns are first analyzed. Wandering spur patterns can be categorized into three cases based on the input to the MASH 1-1-1 DDSM and the detailed analysis for each case is given. The MASH 2-1 DDSM is also analyzed to illustrate how the analysis method can be applied to other DDSM structures.

5.1 Structures of MASH 1-1-1 Internal States and Wandering Spurs

In Chap. 4, it was shown that MASH 1-1-1 internal states $y_2[n]$ and $e_2[n]$ carry the underlying wandering spur pattern that is ultimately manifest in the output phase noise contribution of the divider controller. In this section, the time-domain waveforms of $y_2[n]$ and $e_2[n]$ in the small input case are studied first; their spectral properties will be investigated next.

5.1.1 Parabolic Structure of $e_2[n]$ and the Chirping in $y_2[n]$

The block diagram of the MASH 1-1-1 DDSM divider controller considered is reproduced in Fig. 5.1. A 1-bit LSB dither $d[n]$ is applied at the input of the second stage EFM1.

D. Mai, M. P. Kennedy, *Wandering Spurs in MASH-based Fractional-N Frequency Synthesizers*, Analog Circuits and Signal Processing, https://doi.org/10.1007/978-3-030-91285-7_5

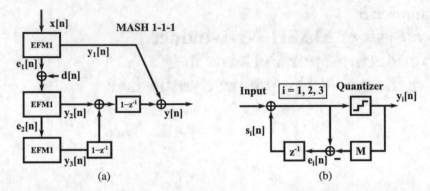

(a)   (b)

Fig. 5.1 Block diagrams of (**a**) a MASH 1-1-1 and (**b**) a first-order error feedback modulator (EFM1). For a conventional LSB-dithered MASH 1-1-1, a 1-bit dither $d[n]$ is introduced to break up the cycles in the quantization error

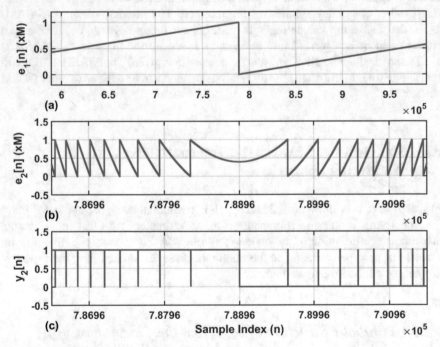

Fig. 5.2 Waveforms of (**a**) $e_1[n]$, (**b**) $e_2[n]$, and (**c**) $y_2[n]$. Input $X = 3$ and $M = 2^{20}$

Consider the simplest case where the input X is close to zero. Example waveforms of $e_1[n]$, $e_2[n]$, and $y_2[n]$ are shown in Fig. 5.2.

Define

$$e_2[n] = s_2[n + 1] = \big(F_{e_2}[n]\big) \bmod M, \tag{5.1}$$

where

$$F_{e_2}[n] = s_2[0] + \sum_{k=0}^{n} (s_1[0] + d[k] + (k+1)X). \qquad (5.2)$$

Note the parabolic shape in Fig. 5.2b. Due to the modulo operation, a parabolic shape will appear in $e_2[n]$ when n_0 satisfies [13]:

$$F_{e_2}[n_0 + 1] - F_{e_2}[n_0] \approx X + e_1[n_0] + \mathbf{E}(d[n]) \approx kM \equiv 0 \ (\mathrm{mod}\ M), \qquad (5.3)$$

where k is an integer and $F_{e_2}[n]$ is as defined in (5.2). The expression for $e_2[n]$ around n_0 can be written as

$$e_2[n] = \left(e_2[n_0] + \sum_{k=n_0+1}^{n} (e_1[k] + d[k]) \right) \mathrm{mod}\ M \qquad (5.4)$$

$$\approx \left(e_2[n_0] + (n - n_0)(e_1[n_0] + \mathbf{E}(d[n])) + \frac{(n - n_0)(n - n_0 + 1)}{2} X \right) \mathrm{mod}\ M, \qquad (5.5)$$

The output $y_2[n]$ is a sequence of zeros and ones. When the value of the expression under the modulo operation in (5.5) reaches multiples of M, changes in the parity in $y_2[n]$ will occur. The indices of these parity changes are solutions of

$$e_2[n_0] + (n - n_0)(-X) + \frac{(n - n_0)(n - n_0 + 1)}{2} X = kM, \ k = 1, 2, \ldots \qquad (5.6)$$

Equation (5.3) is used here. These solutions n_k are given by

$$n_k = n_0 + \frac{1}{X} \left(-\left(\frac{X}{2} + e_1[n_0] + \mathbf{E}(d[n]) \right) \right.$$

$$\left. \pm \sqrt{\left(\frac{X}{2} + e_1[n_0] + \mathbf{E}(d[n]) \right)^2 - 2X (e_2[n_0] - kM)} \right). \qquad (5.7)$$

Note that, because the indices are integers, the solutions n_k should be rounded up to the next integer when used for indices. Here n_k also corresponds to the index of the kth parity change in the $y_2[n]$ sequence around n_0, by definition. In the vicinity of n_0 (for example, $n_0 = 7.8896 \times 10^5$ in Fig. 5.2), when $n < n_0$, $e_1[n]$ is just less than M and when $n > n_0$, $e_1[n]$ is close to zero, which results in the $e_2[n]$ and $y_2[n]$ waveforms shown in Fig. 5.2.

As the simulated spectrogram of $y_2[n]$ in the small input case from Chap. 4 suggests, the sequence contains components of linearly varying frequency in the vicinity of a wandering spur event. The waveform of $y_2[n]$ is a train of unit impulses.

A discrete-time impulse train with linearly varying frequency can be regarded as a sampled version of a continuous chirp signal. The continuous chirp signal should have a fundamental frequency that has the form

$$f(t) = f_0 + (t - t_0)\alpha', \tag{5.8}$$

where f_0 is the frequency of the fundamental component at time t_0 and α' is the rate of change of the frequency. The phase increment of the waveform, or the phase increment of the fundamental component, is then

$$\Delta\phi(t_0, t) = 2\pi \int_{t_0}^{t} \left(f_0 + (t - t_0)\alpha' \right) dt = (t - t_0) f_0 + \frac{1}{2}\alpha'(t - t_0)^2. \tag{5.9}$$

Define the sampling frequency as T_s, let $t_0 = T_s n_0$ and $t = T_s n$, then the discrete phase increment of the component at the fundamental frequency is

$$\Delta\phi[n_0, n] = \Delta\phi(t_0, t) = 2\pi \left((n - n_0) F_0 + \frac{1}{2}\alpha(n - n_0)^2 \right) \tag{5.10}$$

where $F_0 = T_s f_0$ and $\alpha = T_s^2 \alpha'$. Since $F_{e_2}[n]$ is related to the timing of overflows, i.e., the phase of $e_2[n]$ and $y_2[n]$, and it has a structure identical to (5.10), the waveforms of $e_2[n]$ and $y_2[n]$ should exhibit a linear chirp when a parabolic shape in $e_2[n]$ is present in the small input case. If the sampling frequency is sufficiently high, the corresponding discrete signal should have a fundamental frequency

$$F[n] = T_s f(T_s n) = F_0 + (n - n_0)\alpha. \tag{5.11}$$

Assume that a unit impulse appears in the discrete impulse train at $n = n_0$. The next unit impulse happens at $n = n_0 + N$ when the phase increment is $l2\pi$, where l is an integer. Therefore,

$$2\pi \left(N F_0 + \frac{1}{2}\alpha N^2 \right) = l2\pi, \tag{5.12}$$

$$F\left[n + \frac{N}{2}\right] = F_0 + \frac{1}{2}\alpha N = \frac{l}{N} = \frac{1}{N/l}. \tag{5.13}$$

Equation (5.13) suggest that the average frequency found between two impulses corresponds to the instantaneous frequency at the midpoint between the impulses when the impulse train has a linearly varying frequency. Specifically, when $l = 1$,

$$F\left[n + \frac{N}{2}\right] = F_0 + \frac{1}{2}\alpha N = \frac{1}{N}. \tag{5.14}$$

This indicates that the frequency estimated from the adjacent impulses, or ones in the waveform, should be equal to the instantaneous frequency of the time-varying fundamental component at the midpoint between them if the impulse train has a linearly varying frequency. Next, the waveform of $y_2[n]$ in the small input case is analyzed with knowledge of this property.

When $n > n_0$, n_k corresponds to an occurrence of one and the period of $y_2[n]$ between adjacent parity alternations in the $y_2[n]$ sequence is therefore

$$N_{k,k-1} = n_k - n_{k-1}.$$ (5.15)

The rate of change of the frequency of the appearance of ones can be estimated to be

$$\frac{2}{(n_{k+1} - n_{k-1})} \left(\frac{1}{N_{k+1,k}} - \frac{1}{N_{k,k-1}} \right) = \frac{X}{M},$$ (5.16)

where the first term is the reciprocal of the number of sampling periods between the centers of two adjacent periods of ones and the second term is the difference between the frequencies estimated from the two adjacent periods of ones.

Equation (5.16) indicates that a linearly increasing frequency component exists in the $y_2[n]$ sequence when $n > n_0$. When $n < n_0$, n_k corresponds to an occurrence of zero and a similar analysis applies.

The rate of change of frequency corresponds to the spur *wander rate*. The wander rate associated with a single parabolic component is the slowest and the most prominent; we call this the *fundamental wander rate*. Due to the symmetry of the quadratic equation's solutions, waveforms of $e_2[n]$ and $y_2[n]$ with linearly *decreasing* frequency can be observed when $n < n_0$; a linearly *increasing* frequency is seen when $n > n_0$, as shown in Fig. 5.2. This corresponds to the wandering spur event at the fundamental wander rate around n_0 in the spectrograms of $y_2[n]$ and $e_2[n]$.

5.1.2 Spectral Properties of $y_2[n]$ and $e_2[n]$

We next investigate the spectral properties of $y_2[n]$. Assume that a single slice spectral analysis in the spectrogram of $y_2[n]$ (e.g. in Fig. 4.5) is found by applying a window $w[n]$ of length N_w which has the Fourier transform $W(\Omega)$. When $n > n_0$, as shown in Fig. 5.3, the $y_2[n]$ sequence in the window, denoted $\hat{y}_2[n]$, is

$$\hat{y}_2[n] = (y_2[n]w[n - \gamma]) = \left(\sum_{l=0}^{l_m-1} \delta \left[n - \sigma - \sum_{p=1}^{l} N_p \right] \right) w[n - \gamma]$$ (5.17)

$$\approx \left(\sum_{l=-\infty}^{\infty} \delta \left[n - \sigma - l N_{avg} \right] \right) w[n - \gamma].$$ (5.18)

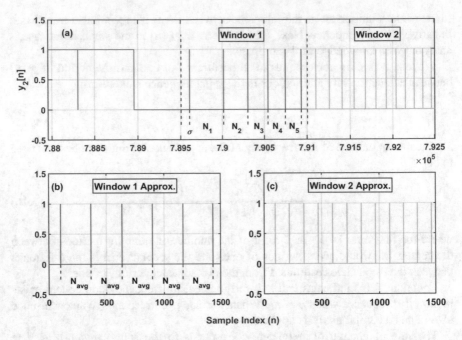

Fig. 5.3 Waveforms of (**a**) $y_2[n]$ of a MASH 1-1-1, (**b**) the approximation in the first window, and (**c**) the approximation in the second window. Input $X = 3$ and $M = 2^{20}$

Equation (5.18) approximates the impulses in the window with a segment of an impulse train with a constant average frequency observed in the window.

The Fourier transform of the sequence in the window is

$$\hat{Y}_2(j\Omega) = \sum_{n=-\infty}^{\infty} \left(\hat{y}_2[n] * \delta[n + \gamma] \right) e^{-j\Omega n} \tag{5.19}$$

$$\approx \frac{1}{N_{avg}} \int_{\Theta=<2\pi>} \left(W(j(\Omega - \Theta)) e^{-j\Theta(\sigma-\gamma)} \sum_{l=-\infty}^{\infty} \delta\left(\Theta - l\frac{2\pi}{N_{avg}} \right) \right) d\Theta. \tag{5.20}$$

Here, σ is the sample index of the first one from the beginning of the window and N_p is the sample number from the pth to the $(p + 1)^{th}$ one in the segment starting from γ in $y_2[n]$. The number of ones in $\hat{y}_2[n]$ is l_m, N_{avg} is the average period, and the asterisk denotes convolution.

In (5.20), we approximated the segment of the $y_2[n]$ sequence by a constant frequency sequence. Equation (5.20) shows that the spectrum of $\hat{y}_2[n]$ will appear to have harmonics. For $n < n_0$, a similar result holds. It should be noted that a component close to DC will show up when n is close to n_0 due to the symmetry of the $y_2[n]$ waveform.

As the waveform of $\hat{y}_2[n]$ evolves with n_k, $1/N_p$ and $1/N_{avg}$ will increase or decrease linearly, as suggested by the previous analysis. Because harmonics exist, multiple traces will be visible in the spectrogram. As the spectrogram shown in Fig. 4.5a suggests, the amplitudes of the harmonic traces decay with their order and this is typical.

Note that

$$y_2[n] \approx \frac{1}{M} \left(e_1[n] - (e_2[n] - e_2[n-1]) \right) \tag{5.21}$$

and $e_1[n]$ does not contain a component of varying frequency. The high-pass filtered $e_2[n]$, shown in Fig. 5.4d, is the main contributor to the wandering spurs. In the

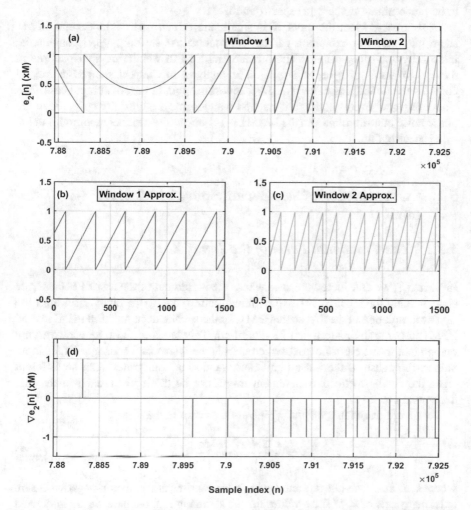

Fig. 5.4 Waveforms of (a) $e_2[n]$, (b) the approximation in the first window, (c) the approximation in the second window, and (d) $\nabla e_2[n]$. Input $X = 3$ and $M = 2^{20}$. Here $\nabla e_2[n] = e_2[n] - e_2[n-1]$

wandering spur events described by (5.20), by adopting a sawtooth approximation to the $\hat{e}_2[n]$ waveform shown in Fig. 5.4b and c, one could perform spectral analysis on $e_2[n]$ and this gives

$$\hat{E}_2(j\Omega) \approx \frac{1}{2\pi} \int_{\Theta=<2\pi>} \left(W(j(\Omega-\Theta))e^{-j\Theta(\sigma-\gamma)} F_{saw,N_{avg}}(j\Theta) \right) d\Theta,$$

(5.22)

where $F_{saw,N_{avg}}(j\Omega)$ is the Fourier transform of the approximated periodic saw-tooth waveform with a period of N_{avg}. Since the amplitudes of the harmonics of a sawtooth waveform decay with the order, the slower moving wandering spurs tend to be predominant in the $e_2[n]$ spectrogram.

In summary, when the input X is small, the error $e_2[n]$ of the second EFM1 stage intermittently exhibits a parabolic structure. As a result, $y_2[n]$ contains an intermittent chirping pattern. These behaviors lead to wandering spur patterns in the spectrograms of $e_2[n]$ and $y_2[n]$. The analysis in Chap. 4 showed that $e_2[n]$ contains the underlying pattern of the wandering spurs in the small input case. The wandering spur events at the fundamental wander rate are typically observed in the phase noise contribution from the MASH 1-1-1 due to their higher amplitudes in the $e_2[n]$ spectrogram.

5.2   Categorization of Wandering Spurs by MASH 1-1-1 Input

5.2.1   The Fractional Form of the Input X

In Sect. 5.1, we considered the case where the increment of the first EFM1 internal state, namely the input X, is small. A linear pattern due to the input arises after the first accumulation in the first stage EFM1. In the second accumulation in the second EFM1, a parabolic pattern can be identified. This leads to the frequency-varying component in the EFM1 output which causes the wandering spur. For larger input values, the cause of the most significant wandering spur pattern can be different since the smallest effective increment might not be the input X itself in the first accumulation.

For an arbitrary integer input X, it can be written in the form

$$X = \frac{kM+r}{D} = \frac{kM}{D} + X',$$

(5.23)

where k, D and r are integers and k and D are coprime and both non-negative. There exist many sets of k, D and r values for the same input X but here we are interested in the value of D that gives the smallest $|r|$. Also, the ratio $X' = r/D$, which is the average increment every D samples during the first accumulation, should lead to a significant parabolic pattern when the input X is accumulated twice. The value of r can be found by evaluating

$$r = (DX - kM), \quad k = 0, \ 1, \ 2, \ldots \tag{5.24}$$

$$= \begin{cases} (DX) \bmod M, & \text{if } (DX) \bmod M < M/2, \\ (DX) \bmod M - M, & \text{otherwise.} \end{cases} \tag{5.25}$$

When $D = 1$, according to (5.23) and (5.25), the increment in the first accumulation is $X' = r$. The small input X case analyzed is such an example. In order to ensure that the period of the wandering spur events is shorter than the observation window length N_w, we require that

$$|r| < r_{max} = M/N_w. \tag{5.26}$$

Here, the window length is defined relative to the phase detector frequency to match the notation in the previous section. Thus $N_w = T_M/T_{PFD}$, where T_M is the length of the window in time used in the output phase noise measurement.

When $D > 1$, D linear traces with offsets of M/D exist in the waveform of $e_1[n]$, since k and D are coprime, by definition. The points of the $e_1[n]$ waveform appear on a linear trace every D samples. Due to equal offsets between adjacent traces, the rate of occurrences of the wandering spur events at the fundamental wander rate is D times that of the case where $D = 1$, assuming equal X'. In other words, the period of the fundamental wandering spur events is related to $|r| = |DX'|$ instead of $|X'|$. Thus, the result in (5.26) still applies in cases where $D > 1$.

In the cases where $k > 0$, the kM/D term in (5.23) should have a significantly smaller period and it consequently causes an insignificant pattern. We require that

$$\frac{\min(k)M}{D} \geq \frac{M}{N_w}. \tag{5.27}$$

Since $\min(k) = 1$,

$$D_{max} = N_w. \tag{5.28}$$

Alternatively, a D_{max} value less than N_w can be chosen since it also satisfies inequality (5.27). For example, for convenience, we could constrain D_{max} to be an exponent of two. The algorithm for putting the input X into the form shown in (5.23) is as follows:

- Find the maximum of the denominator D using (5.28),
- Search among all possible values of D for the one that gives the minimum value of $|r|$,
- Solve for k using (5.23).

If $|r|$ is found to be greater than r_{max} given by (5.26), then the most significant wandering spur pattern will not be visible in the spectrogram.

5.2.2 Categories of Wandering Spurs

With the input in the form of (5.23), we can proceed to categorize wandering spur patterns. When

$$X \le \frac{M}{D_{max} + 1}, \tag{5.29}$$

we have $k = 0$ and $D = 1$; consequently $X = X' = r$. It has been shown in Sect. 5.1 that the $e_2[n]$ waveform periodically exhibits a parabolic shape with a period of M/X. When the input X is small, the window of the spectrum analysis captures the varying frequency component as it moves. However, when X increases, the parabolic shape might occur multiple times within the window. The critical point happens when the period of the occurrence matches the window length N_w. The period of the occurrence will be smaller than the window length if $M/X < N_w$. Consequently, the wandering spur pattern is only apparent when $X < M/N_w$.

When $X > M/D_{max}$, the value of k is greater than zero. Consider a case where $k > 0$ and $r \ne 0$. The ratio of the modulus M to the input X is:

$$\frac{M}{X} = \frac{D}{k + r/M}. \tag{5.30}$$

After accumulating the input X for D times, the change under the modulo operation is r. Therefore, the average rate of change as the result of accumulation under the modulus M is X'. The ratio of the modulus M to $|X'|$ is then

$$\frac{M}{|X'|} = \frac{D}{|r|/M}. \tag{5.31}$$

Because we assume $k > 0$, $M/X < M/|X'|$. This means that the period of the accumulation of X' is longer than the accumulation of X under the modulo operation, giving a more obvious linear pattern that results in a parabolic pattern when accumulated again.

However, when $r = 0$, no change will take place after accumulating the input X for D times, i.e., no apparent linear pattern will appear when X is accumulated.

In summary, the wandering spurs caused by the MASH DDSM can be categorized by the fractional form of the input X. When the input X is written in the form shown in (5.23), three cases exist:

1. Case I: when $k = 0$, i.e., the small input case where the accumulation of the input X leads to a simple linear pattern,

2. Case II: when $k > 0$ and $r \neq 0$, the residue r under the modulo operation is non-zero and produces the most prominent linear pattern after accumulation,
3. Case III: when $k > 0$ and $r = 0$, no residue results under the modulo operation, and the apparent linear pattern does not appear after the first accumulation.

In the first two cases, the wandering spur patterns have a well-defined periodic pattern. In the third case, the pattern of wandering spurs involves the initial condition. The analyses for all three cases of wandering spur patterns will be presented in the following sections.

5.3 Wandering Spur Pattern in Case I: Small Input X Contributes to the Linear Pattern After First Accumulation

Following the analysis of the small input case in Sect. 5.1 and with (5.16), the fundamental wander rate in this case is:

$$WR = f_{PFD}^2 \frac{X}{M}, \tag{5.32}$$

where f_{PFD} is the phase frequency detector update frequency.

The major wandering spur events at the fundamental wander rate take place at the solutions of (5.3), namely

$$n_0 \approx \frac{lM - (s_1[0] + \mathbf{E}(d[n]))}{X}, \quad l = 1, 2, 3, \ldots, \tag{5.33}$$

where $s_1[0]$ is the initial condition of the first EFM1 in the MASH 1-1-1. The period of the major events is thus

$$T_{WS} = \frac{1}{f_{PFD}} \frac{M}{X}. \tag{5.34}$$

In this chapter, the closed-loop model described in Chap. 2 with parameters in Table 2.1 is used to evaluate the contribution of the divider controller to the output phase noise. The spectrograms of $e_2[n]/M$ and the contribution of the MASH 1-1-1 to the output phase noise in the $X = 3$ case from *behavioral simulations* are shown in Fig. 5.5a and b respectively with the estimated wander rate and period highlighted.

Notice that around

$$n_0 = \frac{lM - h(s_1[0] + \mathbf{E}(d[n]))}{hX}, \quad l = 1, 2, 3, \ldots, \tag{5.35}$$

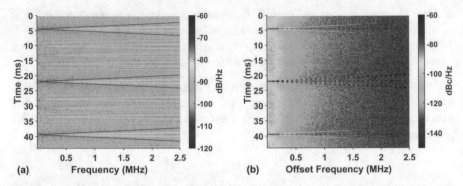

Fig. 5.5 Spectrograms of (**a**) $e_2[n]/M$ and (**b**) output phase noise contribution of the MASH 1-1-1 in the closed-loop behavioral simulation. Input $X = 3$, $M = 2^{20}$ and PFD update frequency $f_{PFD} = 20$ MHz. The black and red dashed lines represent the estimated wander rate and occurrences

the events with higher wander rates on the spectrogram of $e_2[n]/M$ predicted at h times the fundamental wander rate are not observed in the output phase noise spectrogram.

5.4   Wandering Spur Patterns in Case II: Residue Under Modulo Operation Contributes to the Linear Pattern After the First Accumulation

In this section, we focus on the second type of wandering spur. In this case, the input can be written as (5.23) with $k > 0$ and $r \neq 0$. The accumulation of r under the modulo operation leads to a more prominent linear pattern. As discussed in Sect. 4.3, the most prominent wandering spur pattern is at the fundamental wander rate, i.e. the lowest wander rate. In this section, we first show that the residue r will cause a parabolic shape after the second accumulation and the fundamental wander rate will be derived. Then it will be shown that the patterns given by inputs with different D parities are different but the period of the occurrence of the events at the fundamental wander rate can be found using the same equation, regardless of the parity of D.

5.4.1   Time Domain Waveforms and the Fundamental Wander Rate

First rewrite the expression for $e_2[n]$ as

$$e_2[n] \approx \Big(s_2[0] + T_1[n] + T_2[n] \Big) \bmod M, \qquad (5.36)$$

where

$$T_1[n] = \frac{(n+1)(n+2)}{2D} kM, \tag{5.37}$$

$$T_2[n] = \frac{(n+1)(n+2)}{2} X' + (n+1)(s_1[0] + \mathbf{E}(d[n])). \tag{5.38}$$

Assume that $T_1[n]$ in (5.36) alone leads to a period which is denoted Δ. Then one can solve for the values of Δ as follows:

$$T_1[n+\Delta] - T_1[n] = \frac{\Delta(2n+3+\Delta)k}{2D} M = lM, \tag{5.39}$$

where l is an integer. Judging the parity of the numerator of (5.39), in the case of odd D, the minimum solution is $\Delta = D$; for even D, the minimum solution is $\Delta = 2D$. In the following discussion in this section, the minimum Δ is considered unless mentioned otherwise.

If the rest of the n-related terms in (5.36) have a period equal to the minimum period found by (5.39) or multiples of it under the modulo operation, then the $e_2[n]$ waveform temporarily has a steady pattern that corresponds to the vertex of a parabolic shape. This gives

$$T_2[n_0 + \Delta] - T_2[n_0] = lM. \tag{5.40}$$

The solution is

$$n_0 \approx \frac{lM - \Delta(s_1[0] + \mathbf{E}(d[n]))}{\Delta X'}. \tag{5.41}$$

Here, Δ is equal to D or $2D$. Typical waveforms of $e_1[n]$ and $e_2[n]$ around a vertex of the parabolic shape in $e_2[n]$ are shown in Fig. 5.6 for the case $X = M/4 + 1$.

Since $T_1[n]$ has a period of Δ under the modulo operation, the points on the $e_2[n]$ waveform will fall on a parabola every Δ samples around n_0. In this case, a steady pattern is given by the T_1 term. The T_2 term leads to the gradual change in the $e_2[n]$ pattern and, when a parabolic component reaches zero or M, an abrupt change of M or $-M$ in $e_2[n]$ results; this leads to the corresponding parity change in $y_2[n]$. Following (5.5) in Sect. 5.1, the instant of a change of M or $-M$ in $e_2[n]$ around n_0 can be found by solving:

$$P_{e_2[n_0]}[n] = e_2[n_0] - (n - n_0)\frac{(\Delta + 1)}{2} X' + \frac{(n - n_0)(n - n_0 + 1)}{2} X' = lM \tag{5.42}$$

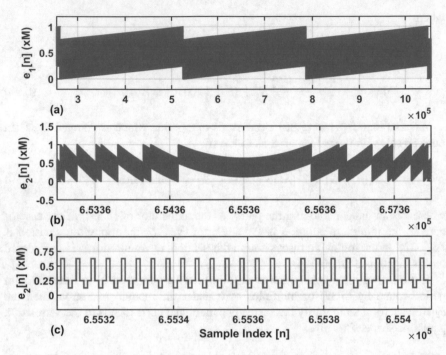

Fig. 5.6 Waveforms of (**a**) $e_1[n]$, (**b**) $e_2[n]$, and (**c**) zoom of $e_2[n]$ around the vertex of the parabola shape where the $e_2[n]$ values are not equally-spaced in amplitude. Input $X = M/4 + 1$ and $M = 2^{20}$. Note the linear pattern in $e_1[n]$ and the parabolic pattern in $e_2[n]$

since we assumed

$$
F_{e_2}[n_0 + \Delta] - F_{e_2}[n_0]
$$
$$
\approx \frac{\Delta(\Delta + 1)}{2} X' + \Delta \left(e_1[n_0] + \mathbf{E}(d[n]) \right) \equiv 0 \ (\mathrm{mod} \ M) \tag{5.43}
$$

and thus the equivalent increment of $\Delta \left(e_1[n_0] + \mathbf{E}(d[n]) \right)$ is $-X'\Delta(\Delta + 1)/2$.

The solutions of (5.42) are

$$
n_{e_2[n_0], l} = n_0 + \frac{1}{X'} \left(\frac{X'\Delta}{2} \pm \sqrt{\left(\frac{X'\Delta}{2} \right)^2 - 2X' \left(e_2[n_0] - lM \right)} \right), \tag{5.44}
$$

where l is an integer. The results should be rounded up when used for indices.

A sawtooth approximation can be applied to a parabolic component of $e_2[n]$. As (5.36) indicates, the parabolic component sequence only takes non-zero values every Δ samples due to the modulo operation. This means that the approximated sawtooth is re-sampled every Δ samples. A parabolic component can be approximated by a sawtooth waveform with a period of N_{avg}, as shown in Fig. 5.7. Denote the single

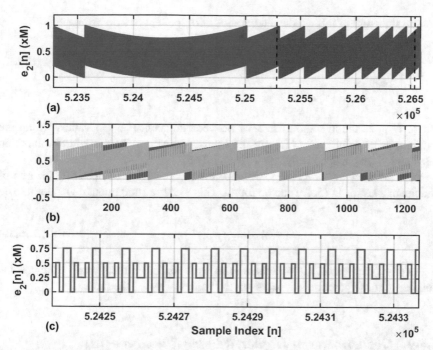

Fig. 5.7 Waveforms of (**a**) $e_2[n]$, (**b**) the approximation by (5.46) in the window, and (**c**) zoomed waveform around the vertex of the parabolic shape where $e_2[n]$ values are equally-spaced. The input is $X = M/4 + 1$ and $M = 2^{20}$

non-repetitive triangular sawtooth waveform starting from zero as $f_{tri}[n]$. Then a parabolic component in $e_2[n]$ can be approximated as:

$$\left(\left(P_{e_2[n_0]}[n] \right) \bmod M \right) w[n - \gamma] \tag{5.45}$$

$$\approx \left(f_{tri}[n] * \sum_l \delta[n - \alpha_{e_2[n_0]} - l N_{avg}] \right) \left(\sum_p \delta[n - n_0 - p\Delta] \right) w[n - \gamma] \tag{5.46}$$

$$\approx \left(f_{tri}[n] \sum_p \delta[n - p\Delta] \right) * \left(\sum_l \delta[n - n_{e_2[n_0],l}] \right) w[n - \gamma] \tag{5.47}$$

$$= f_{e_2}[n] * \left(\sum_l \delta[n - n_{e_2[n_0],l}] \right) w[n - \gamma], \tag{5.48}$$

where $f_{e_2}[n]$ is the non-repetitive re-sampled triangular approximation for the parabolic component with offset $e_2[n_0]$ in the section of $e_2[n]$ within the spectral analysis window $w[n - \gamma]$; the asterisk denotes convolution.

The spectrum of the windowed parabolic component can be estimated as

$$\frac{1}{N_{avg}} \int\limits_{\Theta=<2\pi>} \left(W\left(j(\Omega-\Theta)\right) e^{-j\Theta(\sigma-\gamma)} F_{e_2}(j\Theta) \sum_l \delta\left(\Theta - \frac{2\pi l}{N_{avg}}\right) \right) d\Theta,$$

(5.49)

where N_{avg} is the average difference between adjacent $n_{e_2[n_0],l}$ values within the window, $W(j\Omega)$ and $F_{e_2}(j\Omega)$ are the Fourier transform of the window function $w[n]$ and $f_{e_2}[n]$, respectively, and σ is an offset.

With the analysis of $n_{e_2[n_0],l}$, which is similar to (5.16) in Sect. 5.1, the rate of change of $1/N_{avg}$ in (5.49) can be found. This yields a fundamental wander rate of

$$WR = f_{PFD}^2 \frac{|X'|}{M}.$$

(5.50)

As (5.42) indicates, multiple parabolas with vertical offsets could exist and the number of these parabolas depends on the values of $e_2[n]$ around n_0.

5.4.2 Occurrences of Higher Wander Rates on $e_2[n]$ Spectrogram

If $e_2[n]$ has *equally-spaced* values around n_0, i.e. for each $e_2[n]$ there exists a $c_{n,k}$ for a constant integer h_k such that

$$e_2[n] - e_2[n + c_{n,k}] \approx M/h_k \pmod{M}$$

(5.51)

and define $h = \max(\{h_k\})$, then the perturbations introduced by the parabolic shape happen with a linearly changing frequency because of the linear change introduced by the e_2 term in (5.44). The approximation for the segment of $e_2[n]$ within the window can be written as

$$\hat{e}_2[n] \approx w[n-\gamma] \sum_{m=0}^{\Delta-1} f_{e_2}[n] * \left(\sum_l \delta[n - n_{e_2[n_0+m],l}] \right)$$

(5.52)

$$= w[n-\gamma] \sum_i \sum_{m\in M_i} f_{e_2}[n] * \left(\sum_l \delta[n - n_{e_2[n_0+m],l}] \right).$$

(5.53)

The set M_i contains h different m values such that $e_2[n_0 + m]$ takes on all h different values. In the cases considered in this section, m from 0 to $(\Delta - 1)$

can be divided into an integer number of such sets. Approximating the $n_{e_2[n_0+m],l}$ sequences by arithmetic sequences, we have

$$\hat{e}_2[n] \approx w[n-\gamma] \sum_i f_{e_2}[n] * \left(\sum_l \delta[n - \sigma_i - lN'_{avg}] \right). \tag{5.54}$$

The Fourier transform of $\hat{e}_2[n]$ can be approximated by

$$\hat{E}_2(j\Omega) = \sum_{n=-\infty}^{\infty} \left(\hat{e}_2[n] * \delta[n+\gamma] \right) e^{-j\Omega n} \tag{5.55}$$

$$\approx \sum_i \frac{1}{N'_{avg}} \int_{\Theta=<2\pi>} \left(W(j(\Omega-\Theta)) e^{-j\Theta(\sigma_i-\gamma)} F_{e_2}(j\Theta) \sum_l \delta\left(\Theta - \frac{2\pi l}{N'_{avg}} \right) \right) d\Theta. \tag{5.56}$$

Since $e_2[n_0 + m]$ are equally-spaced and non-repetitive when $m \in M_i$, the terms under the square root in (5.44) are also spaced equally by M/h for the same l. Labeling all $n_{e_2[n_0+m],l}$ in order as $n_1, n_2, \cdots, n_k, \cdots$, N_{avg} is the average of the difference $n_k - n_{k-1}$ within the given window and the rate of change of $1/N'_{avg}$ can be found:

$$\left| \frac{2}{n_{k+1} - n_{k-1}} \left(\frac{1}{n_{k+1} - n_k} - \frac{1}{n_k - n_{k-1}} \right) \right| = \frac{|X'|h}{M}. \tag{5.57}$$

This means that the wander rate is h times the result of (5.50) in this case.

Interestingly, similar occurrences arise in the $M/X > N_w$ cases as well. When $e_1[n_0] \approx kM/h$ and k and h are coprime,

$$e_2[n] \approx \left(e_2[n_0] + \sum_{m=1}^{n-n_0} \frac{kM}{h} + \frac{(n-n_0)(n-n_0+1)}{2} X \right). \tag{5.58}$$

The second term in (5.58) establishes a steady and equally-spaced pattern while the X-related term causes a gradual change with a parabolic shape. Therefore, for these occurrences, a wandering spur with h times the fundamental wander rate can be observed. The wandering spurs with higher wander rates can be seen in the spectrogram of $e_2[n]/M$ in Fig. 5.5a in Sect. 5.3. Equation (5.35) estimates the indices around which these events take place.

According to (5.36), when the T_2 term is contributing almost zero to $e_2[n]$, the T_1 term determines the $e_2[n]$ values around n_0. With different parities of D, it might end up with different wandering spur patterns due to the structure of $T_1[n]$.

5.4.3 $T_1[n]$-Induced Pattern and the Parity of D

In this subsection, we discuss how the $T_1[n]$-induced pattern depends on the parity of D and how it affects the $e_2[n]$ waveform.

5.4.3.1   Even D

First consider an even D. To facilitate the discussion, we define

$$T_1'[n] = \left(\frac{DT_1[n-1]}{M}\right) \bmod D \tag{5.59}$$

and first investigate the structure of $T_1'[n]$.

Theorem 5.1 *The $T_1'[n]$ sequence is symmetrical within a $2D$ period.*

Proof First, since

$$\frac{DT_1[2D-p-2]}{M} = \frac{(2D-p-1)(2D-p)k}{2}$$
$$\equiv \frac{p(p+1)k}{2} \pmod{D} = \frac{DT_1[p-1]}{M} \tag{5.60}$$

and, with the fact that a period of the sequence is $2D$, it should be noticed that the $T_1'[n]$ sequence is symmetrical within a $2D$ period, regardless of the parity of D. $\square$

If we had an integer l such that

$$T_1'[n] = \left(\frac{n(n+1)k}{2}\right) \bmod D = l, \tag{5.61}$$

then the value lM/D appears in the $T_1[n]$ sequence. In the case of an even D, we have the following theorem.

Theorem 5.2 *When D is even, $T_1'[n]$ exhibits an equally-spaced pattern with at least $h = 2$.*

The proof of Theorem 5.2 is presented in Appendix B.1.

This results in wandering spur events which occur at at least twice the fundamental wander rate in the spectrogram of $e_2[n]$. In the special case where $D = 2^p$, we have a further conclusion as follows.

Theorem 5.3 *In the cases where $D = 2^p$, $T_1'[n]$ will take all values from 0 to $D-1$, i.e. the waveform of $e_2[n]$ can be found equally-spaced with $h = D$.*

The proof of Theorem 5.3 is presented in Appendix B.2.

Note that $e_2[n]$ can be approximated by

$$e_2[n] \approx \left(e_2[n_0] + (n - n_0)(e_1[n_0] + \mathbf{E}(d[n])) \right.$$

$$\left. + \frac{(n - n_0)(n - n_0 + 1)}{2} \frac{kM}{D} \right) \mod M.$$

$$(5.62)$$

Notice that the last term in (5.62) has the same form as $T_1[n]$. When $e_1[n_0] + \mathbf{E}(d[n]) \approx rM \equiv 0 \pmod{M}$ and r is an integer, the $e_2[n]$ waveform will exhibit a pattern that is determined by the last term in (5.62). In this case,

$$s_1[0] + \mathbf{E}(d[n]) + (n_0 + 1)\left(\frac{kM}{D} + X' \right) \tag{5.63}$$

$$\equiv s_1[0] + \mathbf{E}(d[n]) + (n_0 + 1)X' + \frac{aM}{D} \pmod{M} = rM,$$

$$a = 0, \ 1, \ \cdots D - 1. \tag{5.64}$$

The fact that $nk \mod D$ can take all values from zero to $D - 1$ is used here. Since

$$(n + \delta)k \mod D = (nk \mod D + \delta k \mod D) \mod D, \tag{5.65}$$

$\delta k \mod D \neq 0$ when δ is not an integer times D due to the coprime relationship. For $0 \leq \delta < D$, $(n + \delta)k \mod D$ takes all D values from zero to $D - 1$.
Then

$$n_0 \approx \frac{(Dr - a)M - D(s_1[0] + \mathbf{E}(d[n]))}{DX'} \tag{5.66}$$

$$= \frac{lM - D(s_1[0] + \mathbf{E}(d[n]))}{DX'}. \tag{5.67}$$

It should be noted that the values of n_0 given by (5.67) are a subset of those found by solving (5.41) when $\Delta = 2D$. Equation (5.67) indicates that only half of the solutions given by (5.41) correspond to the events at the fundamental wander rate. The period of the occurrence of the events at the fundamental wander rate in this case is thus

$$2\frac{1}{f_{PFD}}\frac{M}{2D|X'|} = \frac{1}{f_{PFD}}\frac{M}{D|X'|}. \tag{5.68}$$

For the solutions of (5.41) which do not take values of (5.67), the second term in (5.62) disturbs the equally-spaced parabolic pattern; consequently, wandering spur events at the fundamental wander rate will appear in the spectrogram of $e_2[n]$ in those cases. This could be explained by the structure of $e_2[n]$ at those events. When the $e_2[n]$ waveform exhibits a parabolic pattern, a wandering spur event could take place. At the vertex of the parabola in $e_2[n]$ around n_0,

$$e_2[n_0] \approx e_2[n_0 + \Delta]. \tag{5.69}$$

Equation (5.62) indicates that

$$\Delta(e_1[n_0] + \mathbf{E}(d[n])) \approx pM. \tag{5.70}$$

Here p is an integer. Therefore, in the case of an even D,

$$e_1[n_0] + \mathbf{E}(d[n]) \approx e_1[n_0] \approx \frac{pM}{\Delta} = \frac{pM}{2D}. \tag{5.71}$$

The increment of $e_1[n]$ is $X \approx kM/D$ in this case. Due to the coprime relation between k and D,

$$e_1[n'_0] = e_1[n_0 + \delta] \approx \left[\frac{pM}{2D} + \delta\left(\frac{kM}{D}\right) \right] \bmod D = \frac{(p+2l)M}{2D}, \tag{5.72}$$

where l is an integer. Note that when p is even, then $e_1[n'_0] \approx e_1[n'_0] + \mathbf{E}(d[n]) \approx M$ around n_0 and these correspond to the events given by (5.67).

When p is odd, since k is also odd in the even D case, around n_0 we have

$$e_1[n'_0] = e_1[n_0 + \delta] \approx \frac{kM}{2D}, \tag{5.73}$$

Similar to (5.62), one can express $e_2[n]$ with respect to $e_2[n'_0]$ as

$$e_2[n] \approx \left(e_2[n'_0] + (n - n'_0)(e_1[n'_0] + \mathbf{E}(d[n])) \right.$$

$$\left. + \frac{(n - n'_0)(n - n'_0 + 1)}{2} \frac{kM}{D} \right) \bmod M. \tag{5.74}$$

Let $m = n - n'_0 - 1$, the last two terms in (5.74) can be combined as

$$\left((m+2)^2 - 1 \right) \frac{kM}{2D}, \tag{5.75}$$

which is a shifted version of $n^2(kM/(2D))$. This is a different second-order polynomial to the last term in (5.62) that has identical structure to $T_1[n]$, which can be seen as a shifted version of $n(n+1)(kM/(2D))$. Therefore, as shown in Appendix B.4, $e_2[n]$ does not exhibit an equally-spaced pattern in amplitude due to

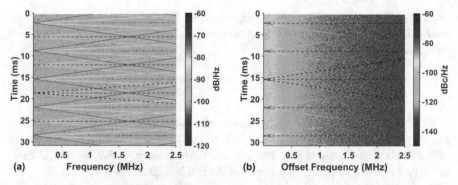

Fig. 5.8 The spectrograms of (a) $e_2[n]/M$ and (b) output phase noise contribution of the MASH 1-1-1 in the closed-loop behavioral simulation. Input $X = \lfloor M/6 + 1 \rceil$, $M = 2^{20}$ and PFD update frequency $f_{PFD} = 20$ MHz. The black and red dashed lines represent the estimated wander rate and occurrences. On the $e_2[n]/M$ spectrogram, an event with twice that of the fundamental wander rate is highlighted; it does not show up in the output phase noise contribution

the different structure of the second-order polynomial under the modulo operation in this case.

The spectrograms of $e_2[n]/M$ and the output phase noise contribution of the MASH 1-1-1 when the input is $X = \lfloor M/6 + 1 \rceil$ are shown in Fig. 5.8a and b. The most significant wandering spur pattern at the fundamental wander rate in the $e_2[n]$ spectrogram is highlighted in the output phase noise contribution of the MASH 1-1-1.

5.4.3.2   Odd D

When D is odd, we have the following theorem.

Theorem 5.4 *For D that is odd and greater than one, $T_1'[n]$ is not equally-spaced i.e. $e_2[n]$ does not have equally-spaced values around n_0 given by (5.41).*

The proof of Theorem 5.4 is presented in Appendix B.3.

Since in this case $e_2[n_0] \approx e_2[n_0 + D]$, around n_0 we have $D(e_1[n_0] + \mathbf{E}(d[n])) \approx pM$, which gives

$$e_1[n_0] + \mathbf{E}(d[n]) \approx e_1[n_0] \approx \frac{pM}{D}. \qquad (5.76)$$

Again due to the coprime relation between k and D, around n_0 there exists $n_0' = n_0 + \delta$ such that

$$e_1[n_0'] = e_1[n_0 + \delta] \approx \frac{kM}{D}. \qquad (5.77)$$

Sequence $e_2[n]$ around n_0' can be expressed as (5.74). Substituting the approximation of $e_1[n_0] + \mathbf{E}(d[n])$ and letting $m = n - n_0' - 1$, the last two terms in (5.74) can be expressed as

$$\frac{(m+2)(m+3)}{2}\frac{kM}{D} = \frac{((m+1)+1)((m+1)+2)}{2}\frac{kM}{D} \tag{5.78}$$

which is a shifted version of $T_1[n]$. Therefore, $e_2[n]$ does not exhibit an equally-spaced pattern around n_0 given by (5.67) in the odd D case. Wandering spur events of the fundamental wander rate take place at these instances. For the odd D case, the period of the wandering spur events at the fundamental wander rate is

$$\frac{1}{f_{PFD}}\frac{M}{D|X'|}. \tag{5.79}$$

For the cases with odd D, Eq. (5.41) gives the indices around which the wandering spur events at the fundamental wander rate happen. An example with $X = \lfloor M/3 + 2 \rceil$ is shown in Fig. 5.9.

As (5.62) suggests, for both even and odd D, if

$$e_1[n_0] = \frac{kM}{N\Delta}, \tag{5.80}$$

where k and N are integers and k and $N\Delta$ are coprime, for every Δ samples, the $e_1[n_0]$-related term in (5.62) will introduce a difference of kM/N. With this difference at every Δ samples, the pattern with a period of Δ is shifted by a multiple of kM/N. Since k and $N\Delta$ are mutually prime, the multiple of kM/N will take values of 0, M/N, $\cdots$, $(N-1)M/N$ due to the modulo operation. As a result, the pattern with a period of Δ from $T_1[n]$ is shifted by equally-spaced values and

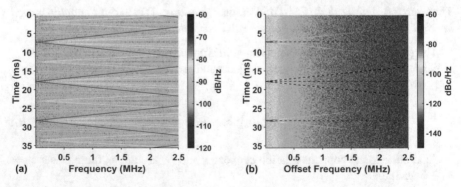

Fig. 5.9 The spectrograms of (**a**) $e_2[n]/M$ and (**b**) output phase noise contribution of the MASH 1-1-1 in the closed-loop behavioral simulation. Input $X = \lfloor M/3 + 2 \rceil$, $M = 2^{20}$ and PFD update frequency $f_{PFD} = 20$ MHz. The black and red dashed lines represent the estimated wander rate and occurrences

this gives wandering spur events at multiples of the fundamental wander rate in the spectrogram of $e_2[n]$. In these cases, $T_2[n + N\Delta] - T_2[n] \approx rM \equiv 0 (\text{mod } M)$, where r is an integer.

Therefore, in general, the indices around which events with multiples of the fundamental wander rate take place can be found:

$$n_0 \approx \frac{lM - N\Delta(s_1[0] + \mathbf{E}(d[n]))}{N\Delta X'}. \tag{5.81}$$

For events at higher wander rates, according to (5.48), the greater Δ values lead to the equivalent down-sampling. Also, N'_{avg} in (5.56) has smaller values compared to N_{avg} in (5.49) in those cases. Therefore, wandering spur events at the fundamental wander rate have the highest amplitudes in the $e_2[n]$ spectrogram. Consequently, they will be most significant in the output phase noise spectrogram in the presence of nonlinearity, as manifested in Fig. 5.9b.

Notice that (5.41) gives the events with $N = 1$ and the minimal Δ. Therefore, these events tend to have the slowest wander rate in the spectrogram of $e_2[n]$.

In conclusion, in both even and odd D cases, the period of the wandering spurs at the fundamental wander rate is

$$T_{WS} = \frac{1}{f_{PFD}} \frac{M}{D|X'|}. \tag{5.82}$$

5.4.4 Summary

The theoretical predictions of the wandering spur wander rates and periods for Cases I and II are summarized in Table 5.1. Note that the predicted wander rates and periods are independent of the nonlinearity and the initial conditions.

Experimental results that confirm the theoretical predictions will be presented in Sect. 5.7.

As outlined in the categorization of wandering spurs in Sect. 5.2, the wandering spurs are visible when the spectral analysis has no or little averaging effect. In Case I and Case II, the averaging effect becomes apparent when the period specified in Table 5.1 is less than the observation window length. In this case, multiple

Table 5.1 Summary of wander rate (fundamental) and period

Input form			
$X = \frac{kM}{D} + X'$			
Input case	Condition	Wander rate	Period
Case I	$k = 0, \; X = X'$	$f_{PFD}^2 \frac{X}{M}$	$\frac{1}{f_{PFD}} \frac{M}{X}$
Case II	$k \neq 0, \; X' \neq 0$	$f_{PFD}^2 \frac{|X'|}{M}$	$\frac{1}{f_{PFD}} \frac{M}{D|X'|}$

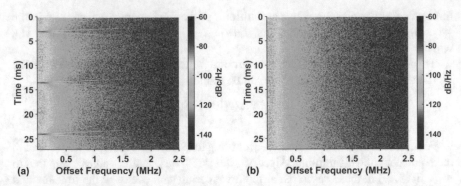

Fig. 5.10 Simulated spectrograms of output phase noise contribution from a MASH 1-1-1 when (**a**) input $X = 5$ and (**b**) input $X = 300$, $M = 2^{20}$ and PFD update frequency $f_{PFD} = 20$ MHz

significant events will take place within the same window and the spectral analysis interprets the wandering spur phenomenon as noise.

Examples of a Case I output phase noise contribution from a MASH 1-1-1 are shown in Fig. 5.10. An observation window of $N_{w,M} = 409600$ points is applied to the output phase noise sampled at a period of $T_{s,M} = 0.5 \times 10^{-9}$ s. The duration of the window is $T_M = N_{w,M} T_{s,M}$. The maximum input for wandering spur observation in Case I, which leads to a wandering spur period that is equal to the observation window duration, is

$$X_{max} = \frac{M}{\dfrac{T_M}{T_{PFD}}} = \frac{M}{N_{w,M} \dfrac{T_{s,M}}{T_{PFD}}} = \frac{2^{20}}{409600 \times \dfrac{0.5 \times 10^{-9}}{1/(20 \times 10^6)}} = 256. \tag{5.83}$$

In Fig. 5.10a, the input to the MASH 1-1-1 is $X = 5$. As the theory predicts, wandering spur events occur at a period of

$$T_{WS} = \frac{1}{f_{PFD}} \frac{M}{X} = \frac{1}{20 \times 10^6} \times \frac{2^{20}}{5} = 10.49 \text{ ms}. \tag{5.84}$$

In Fig. 5.10b, the input is $X = 300$, which is greater than X_{max}. Therefore, no wandering spur is observed.

The observation of wandering spurs in Case I and Case II depends on the ratio of the event period and the observation period. The observation period is relatively constant for certain applications of a fractional-N frequency synthesizer. Furthermore, in Case I and Case II, the period of the spurs is proportional to the modulus of the MASH divider controller. This means that it is more likely to observe the wandering spur events in a fractional-N frequency synthesizer with a MASH DDSM divider controller with a larger modulus since more values of input lead to event periods that are shorter than a given observation period.

5.5   Wandering Spur Patterns in Case III: Large Input X and No Residue Under Modulo Operation After the First Accumulation

In the third case of wandering spur patterns, the MASH 1-1-1 input X can be expressed in the form

$$X = \frac{kM}{D}, \qquad (5.85)$$

i.e. (5.23) with $k > 0, D > 0$, and $r = 0$. In this case, the wandering spur pattern depends *explicitly* on the initial condition of the DDSM, leading to extremely complex behavior.

In Case III, the overflow period in the first EFM1 is short and $e_1[n]$ repeats with identical values in each cycle. The initial state of the first EFM1 leads to the apparent linear pattern in the second EFM1 quantization error $e_2[n]$, which causes the parabolic structure in the signal $e_3[n]$. Therefore, chirp behavior is observed in both $e_3[n]$ and $y_3[n]$.

Figure 5.11 shows the waveforms of the error signals in each EFM1 when $X = M/4$. This is qualitatively different in nature from the Case II $X = M/4 + 1$ example

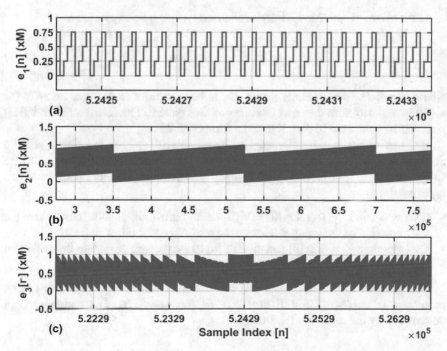

Fig. 5.11 Waveforms of (**a**) $e_1[n]$, (**b**) $e_2[n]$, and (**c**) $e_3[n]$. Input $X = M/4$, $s_1[0] = 1$ and $M = 2^{20}$

shown in Fig. 5.6 in Sect. 5.4, where the linear pattern is exhibited in $e_1[n]$ and the parabolic structure is developed in $e_2[n]$.

We consider the case when $D = 2^a$, and k and D are coprime. The $e_1[n]$ sequence repeats itself with a short period of D:

$$e_1[n] = \left(s_1[0] + (n+1)\frac{kM}{D} \right) \bmod M. \tag{5.86}$$

Furthermore, since

$$(n + \delta)k \bmod D = (nk \bmod D + \delta k \bmod D) \bmod D, \tag{5.87}$$

and $\delta k \bmod D \neq 0$ when δ is not a multiple of D, the value of $nk \bmod D$ does not repeat within D samples. There are only D possible values of $nk \bmod D$ and this means that $e_1[n]$ will contain D different values.

This leads to the expression for $e_2[n]$:

$$e_2[n] = \left(\sum_{m=0}^{n} d[m] + s_2[0] + (n+1)s_1[0] + \sum_{m=0}^{n} \left((m+1)\frac{kM}{D} \right) \right) \bmod M. \tag{5.88}$$

In the above equations, the initial condition $s_1[0]$ can be expressed as

$$s_1[0] = \frac{k'M}{D_{s_1}} + s_1'[0], \tag{5.89}$$

where k' and D_{s_1} are mutually prime. It can be seen from (5.88) that the $s_1[0]$ (or $s_1'[0]$) term determines the most prominent linear pattern. The form can be found by a similar method used to cast X into the form of (5.23).

Because dither is normally applied, even if $s_1'[0] = 0$, the dither will be accumulated to give a linear pattern in $e_2[n]$. The absence of the dither will lead to no wandering spur pattern when $s_1'[0] = 0$. Therefore, wandering spurs of two types exist:

- Case A: when $k' = 0$ and $s_1[0] = s_1'[0]$ i.e. the small initial value case where the accumulation of the input X leads to the most apparent linear pattern,
- Case B: when $k' > 0$, $s_1'[0]$ and dither $d[n]$ cause the most apparent linear pattern after accumulation.

The analysis in [15] indicates that the traces of wandering spurs will not be prominent when $D_{s_1}(s_1'[0] + \mathbf{E}(d[n])) > M/N_w$, where N_w is a window length determined by the reference frequency.

5.5.1 Case A: Small Initial Value Case

First consider the case where the initial condition $s_1[0]$ is small. The quantization error of the third EFM1 $e_3[n]$ is then

$$e_3[n] \approx (s_3[0] + T_3[n] + T_4[n]) \mod M, \tag{5.90}$$

where

$$T_3[n] = \frac{(n+1)(n+2)}{2}(s_1[0] + \mathbf{E}(d[n])) + (n+1)s_2[0], \tag{5.91}$$

$$T_4[n] = \sum_{p=0}^{n}\left(\frac{(p+1)(p+2)}{2}\frac{kM}{D}\right). \tag{5.92}$$

The period of the $T_4[n]$ term is discussed first.

$$T_4[n+\Delta] - T_4[n] \tag{5.93}$$

$$= \frac{\Delta\left(3n^2 + (3\Delta+12)n + (\Delta+1)(\Delta+5)+6\right)}{3}\frac{kM}{2D}. \tag{5.94}$$

The denominator of (5.94) indicates that Δ should be at least $2D$. Since $D = 2^a$ and an integer power of two has a residual of one or two when divided by three, the $(\Delta+1)(\Delta+5)$ term will be divisible by three. Thus, the minimum period is

$$\Delta_{min} = 2D. \tag{5.95}$$

To locate the wandering spur events which occur when the parabolic shape in $e_3[n]$ reaches a vertex, we solve

$$T_3[n_0 + \Delta] - T_3[n_0] = lM. \tag{5.96}$$

The solutions are

$$n_0 \approx \frac{lM - \Delta s_2[0]}{\Delta(s_1[0] + \mathbf{E}(d[n]))}. \tag{5.97}$$

The fundamental wander rate observed in the spectrogram of $e_3[n]$ in this case is

$$WR = f_{PFD}^2\left(\frac{s_1[0] + \mathbf{E}(d[n])}{M}\right). \tag{5.98}$$

In the spectrogram of $e_3[n]$, wandering spur events will happen at the fundamental wander rate or multiples of it.

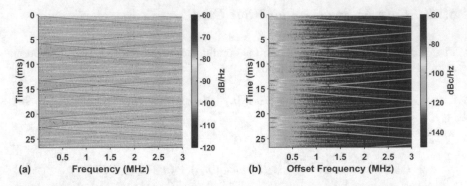

Fig. 5.12 Spectrograms of (**a**) $e_3[n]/M$ and (**b**) output phase noise contribution of the MASH 1-1-1 in the closed-loop behavioral simulation. Input $X = M/4$, $s_1[0] = 3$, $M = 2^{20}$ and PFD update frequency $f_{PFD} = 20$ MHz. One-bit $d[n]$ is generated from the MATLAB uniform random number generator. Note the fixed spur in (b) at an offset of 2.5 MHz

A filtered version of $e_3[n]$, i.e. $\nabla^2 e_3[n] = e_3[n] - 2e_3[n-1] + e_3[n-2]$, contributes to the accumulated quantization error of the MASH 1-1-1 [15]. This leads to the pattern in the spectrogram of the output phase noise of the synthesizer. Since the parabolic shape exists in the quantization error of the *third* EFM1 stage, multiple spurs moving at different rates can be observed, even without the presence of nonlinearity. This is qualitatively different from Cases I and II, which were discussed in Sect. 5.3 and Sect. 5.4, where wandering spurs from the MASH 1-1-1 are not seen at the output of the synthesizer *unless* nonlinearity is present. In the presence of a nonlinearity, some events may become more significant than others in the spectrogram in Case III.

The behavioral model in Chap. 2 with the parameters in Table 2.1 is used for the following simulations. An example with input $X = M/4$, $s_1[0] = 3$, and $M = 2^{20}$ is shown in Fig. 5.12. In this case $\Delta_{min} = 8$. Note that, in the spectrogram shown in Fig. 5.12b, a tone associated with $e_2[n]$ is present at $f_{PFD}/\Delta_{min} = 2.5$ MHz. The frequency of this spurious tone is lower than that of the predicted integer boundary spur fundamental at $X f_{PFD}/M$ [31]. The cause of this spur will be analyzed in the next subsection.

5.5.2   Case B: Large Initial Value Case

Typically, when $(s_1[0] + \mathbf{E}(d[n])) > M/N_w$, the wandering spur pattern does not depend on the initial condition directly. In this case, the initial condition $s_1[0]$ can be put in the form of (5.89). Therefore, the quantization error of the third EFM1 can be expressed as

$$e_3[n] \approx \left(s_3[0] + T_3'[n] + T_4[n] + T_5[n]\right) \bmod M, \tag{5.99}$$

where $T_3'[n]$ is defined by substituting $s_1[0]$ with $s_1'[0]$ in (5.91) and

$$T_5[n] = \frac{(n+1)(n+2)}{2} \frac{k'M}{D_{s_1}}. \tag{5.100}$$

To find the period of $T_5[n]$, one should solve

$$T_5[n+\Delta] - T_5[n] = \frac{\Delta(2n+3+\Delta)k'}{2D_{s_1}} M = lM. \tag{5.101}$$

The minimum solution is $\Delta = D_{s_1}$ when D_{s_1} is odd or $\Delta = 2D_{s_1}$ when D_{s_1} is even. Since $T_4[n]$ has a period of $2D$ or a multiple of $2D$, to find the vertices of the parabolic shapes in $e_3[n]$ in this case, we have

$$T_3'[n_0 + \Delta] - T_3'[n_0] = lM, \tag{5.102}$$

and the corresponding solutions are

$$n_0 \approx \frac{lM - \Delta s_2[0]}{\Delta(s_1'[0] + \mathbf{E}(d[n]))}, \tag{5.103}$$

where the minimum period is

$$\Delta_{min} = LCM(2D_{s_1}, 2D). \tag{5.104}$$

Here LCM denotes the least common multiple.

The fundamental wander rate in this case is

$$WR = f_{PFD}^2 \left(\frac{s_1'[0] + \mathbf{E}(d[n])}{M} \right). \tag{5.105}$$

Wandering spur events at the fundamental wander rate and multiples of it can be observed at the instants that are defined by (5.103). Figure 5.13 shows an example where $X = M/4$, $s_1[0] = \lfloor M/3 + 1 \rfloor$, and $M = 2^{20}$. In this case, $\Delta_{min} = 24$. The output phase noise spectrum also contains a fixed tone with a fundamental frequency of $f_{PFD}/\Delta_{min} = 0.833$ MHz. This is due to the fact that $\Delta = \Delta_{min}$ is also a solution of

$$e_2[n+\Delta] - e_2[n] \approx lM, \quad l = \ldots -1, \, 0, \, 1, \ldots \tag{5.106}$$

in these cases.

The pattern with a period of Δ_{min} in $e_2[n]$ leads to the fixed tone at f_{PFD}/Δ_{min} in the presence of a nonlinearity, as shown in Figs. 5.12b and 5.13b.

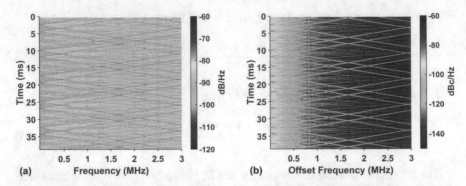

Fig. 5.13 Spectrograms of (**a**) $e_3[n]/M$ and (**b**) output phase noise contribution of the MASH 1-1-1 in the closed-loop behavioral simulation. Input $X = M/4$, $s_1[0] = \lfloor M/3 + 1 \rfloor$, $M = 2^{20}$ and PFD update frequency $f_{PFD} = 20$ MHz. One-bit $d[n]$ is generated from the MATLAB uniform random number generator. Note the fixed spur in (**b**) at an offset of 0.833 MHz

Table 5.2 Summary of wander rate (fundamental) and period

Input/initial condition form
$X = \dfrac{kM}{D} + X', \ \ s_1[0] = \dfrac{k'M}{D_{s_1}} + s_1'[0]$

Condition	Wander rate
$X' = 0, \ k' = 0$	$f_{PFD}^2 \left(\dfrac{s_1[0] + \mathbf{E}(d[n])}{M} \right)$
$X' = 0, \ k' \neq 0$	$f_{PFD}^2 \left(\dfrac{s_1'[0] + \mathbf{E}(d[n])}{M} \right)$

5.5.3 Summary

Table 5.2 summarizes the fundamental wander rates for Case III. Note that wandering spurs are predicted not to appear when $s_1[0] = 0$ and $d[n] = 0$. Measurements that confirm the theoretical predictions will be presented in Sect. 5.7.

Since numerous wandering spur events are present in the spectrograms in Case III, the period of the wandering spur events is not a practical concern in observation. However, considering the nature of a finite-state machine, it can be inferred that the underlying pattern of Case III wandering spurs in a MASH 1-1-1 is periodic. Based on the linear pattern in $e_2[n]$, the period of repetition of the pattern in the $e_3[n]$ spectrogram is

$$T_{Pattern} = \gamma \frac{M}{D_{s_1}(s_1'[0] + \mathbf{E}(d[n]))}, \qquad (5.107)$$

where γ is a factor that depends on the input X and the initial condition $s_1[0]$. From empirical observations, typical values of γ are 0.5, 1, and 2.

5.6 Example of Wandering Spurs in Other DDSMs

In the previous sections, it has been revealed that the indirect or direct contribution from the parabolic internal state of the MASH 1-1-1 causes the wandering spur pattern that is seen in the output phase noise of a fractional-N frequency synthesizer. In fact, wandering spurs may be observed in a fractional-N frequency synthesizer based on any DDSM that is of second or higher order.

The wandering spurs can be traced back to the successive accumulation of a constant within a DDSM. The wandering spurs can be categorized by the first accumulation of the constant; this constant could be $X + E(d[n])$ or $s_1[0] + E(d[n])$, where X is the input, $s_1[0]$ is the initial condition of the first EFM1, and $E(d[n])$ is the expected value of the LSB dither. The most significant linear pattern can be contributed by accumulation of the constant itself, or the residue under the modulo operation. The similar analysis applies to other DDSM structures, e.g., a MASH DDSM with a different modulus and other DDSMs.

Since the wandering spur phenomenon has not been demonstrated in synthesizers based on other divider controllers, an example that illustrates how to apply the analysis method to other divider controller architectures will be presented next. In this section, several different DDSMs that can give rise to wandering spurs will be analyzed.

5.6.1 MASH 2-1 DDSM

The MASH 2-1 has the structure shown in Fig. 5.14. Define the quantization errors of the EFM2 and EFM1 in the MASH 2-1 as $e_1[n]$ and $e_2[n]$, respectively. Different from the EFM1, the EFM2 here has two initial conditions, namely $e_1[-1]$ and

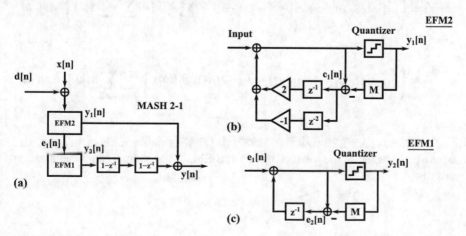

Fig. 5.14 Block diagrams of (**a**) a MASH 2-1, (**b**) a second-order error feedback modulator (EFM2), and (**c**) second-stage EFM1. LSB dither $d[n]$ is applied

$e_1[-2]$. To compare with the MASH 1-1-1 with first-order shaped LSB dither, a dither

$$d[n] = d'[n] - d'[n-1] \qquad (5.108)$$

is considered. Here $d'[n]$ is a sequence of uniformly distributed zeroes and ones and we assume the initial condition $d'[-1] = 0$. Following Appendix C, the expression for $\nabla e_1[n] = e_1[n] - e_1[n-1]$ can be found as

$$(\nabla e_1[n]) \bmod M = \left((e_1[-1] - e_1[-2]) + d'[n] + \sum_{k=0}^{n} x[k] \right) \bmod M \qquad (5.109)$$

$$\approx \left((e_1[-1] - e_1[-2]) + \mathbf{E}(d'[n]) + \sum_{k=0}^{n} x[k] \right) \bmod M. \qquad (5.110)$$

Recall that, in the MASH 1-1-1, the expression for the first EFM1 quantization error $e_1[n]$, when the same dither is applied, is

$$e_1[n] = (e_1[n]) \bmod M = \left(s_1[0] + d'[n] + \sum_{k=0}^{n} x[k] \right) \bmod M. \qquad (5.111)$$

Therefore, comparison leads to the equivalence of

$$s_{1,MASH111}[0] \equiv (e_1[-1] - e_1[-2]). \qquad (5.112)$$

The first stage quantization error of the MASH 2-1 $e_1[n]$ can be expressed as

$$e_1[n] = \left(e_1[-1] + (n+1)(e_1[-1] - e_1[-2]) + \sum_{k=0}^{n} d'[k] + \sum_{p=0}^{n} \sum_{k=0}^{p} x[k] \right) \bmod M \qquad (5.113)$$

$$\approx \left(e_1[-1] + (n+1)(e_1[-1] - e_1[-2] + \mathbf{E}(d'[k])) + \sum_{p=0}^{n} \sum_{k=0}^{p} x[k] \right) \bmod M. \qquad (5.114)$$

For a first-order LSB-dithered MASH 1-1-1 with uniform dither of zeros and ones $d[n]$ injected at the input of the second EFM1, the second stage quantization error is

$$e_2[n] = \left(s_2[0] + \sum_{p=0}^{n} (e_1[p] + d[p]) \right) \bmod M \qquad (5.115)$$

$$= \left(s_2[0] + (n+1)s_1[0] + \sum_{p=0}^{n} d[p] + \sum_{p=0}^{n} \sum_{k=0}^{p} x[k] \right) \bmod M \qquad (5.116)$$

$$\approx \left(s_2[0] + (n+1)(s_1[0] + \mathbf{E}(d[n])) + \sum_{p=0}^{n} \sum_{k=0}^{p} x[k] \right) \bmod M. \qquad (5.117)$$

Comparing (5.114) and (5.117), the equivalence

$$s_{2,MASH111}[0] \equiv e_1[-1] \qquad (5.118)$$

exists between the MASH 2-1 and the MASH 1-1-1. Note that an equivalent double accumulation takes place in the EFM2 of the MASH 2-1, as indicated by (5.114).

A simulation of the phase noise contribution of a MASH 2-1 with input $X = 3M/4 + 1$, $e_1[-1] = 555359$, $e_1[-2] = 524288$, $\mathbf{E}(d'[k]) = 0.5$, $M = 2^{20}$ and $f_{PFD} = 20$ MHz is shown in Fig. 5.15.

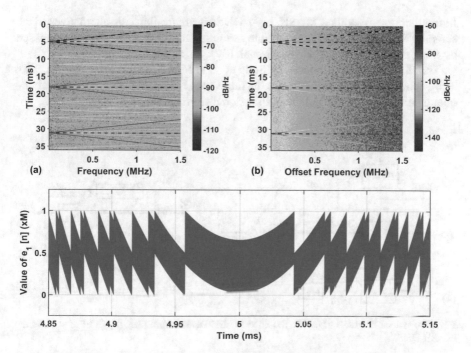

Fig. 5.15 (a) The spectrogram of $e_1[n]/M$, (b) the spectrogram of the output contribution from the MASH 2-1 DDSM, and (c) the waveform of $e_1[n]$, input $X = 3M/4 + 1$, $e_1[-1] = 555359$, $e_1[-2] = 524288$, $\mathbf{E}(d'[k]) = 0.5$, $M = 2^{20}$ and $f_{PFD} = 20$ MHz

The first occurrence of the wandering spur is expected at

$$t = T_{PFD} \frac{M - \Delta(e_1[-1] - e_1[-2] + \mathbf{E}(d'[k]))}{\Delta} \tag{5.119}$$

$$= \frac{1}{20 \times 10^6} \times \frac{M - 2 \times 4 \times (555359 - 524288 + 0.5)}{2 \times 4} \tag{5.120}$$

$$= 5 \text{ ms.} \tag{5.121}$$

This is confirmed by simulation. Note the signature parabolic structure in the $e_1[n]$ waveform of the MASH 2-1 shown in Fig. 5.15c and the underlying pattern of wandering spurs in the spectrogram of $e_1[n]/M$.

Due to the similar structures of the internal states, the MASH 2-1 manifests wandering spurs in an identical manner to the MASH 1-1-1, despite having a more complicated first stage EFM. The double accumulation, or any equivalent of it, within a DDSM-based divider controller is the root cause of wandering spurs.

5.6.2   MASH 1-1-1 Alternatives

In the literature, modified MASH 1-1-1 DDSMs are available as alternatives to the standard MASH 1-1-1 DDSM. Notable examples are the MASH DDSMs proposed by Song and Park [38] and Hosseini et al. [39].

The MASH DDSM proposed by Song et al. has the structure shown in Fig. 5.16. The second and third EFM stages in the MASH structure take the sum of the quantization error signal and the output signal from the previous stage.

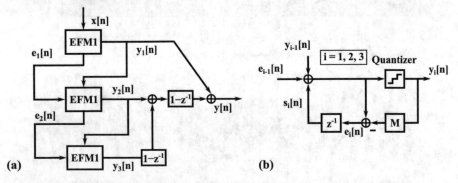

Fig. 5.16 Block diagrams of (**a**) the modified MASH DDSM from [38], (**b**) an EFM stage in the MASH DDSM

The output of each stage can be written as

$$Y_1(z) = \frac{1}{M} \left(X(z) - (1 - z^{-1})E_1(z) \right),$$ (5.122)

$$Y_2(z) = \frac{1}{M} \left(Y_1(z) + E_1(z) - (1 - z^{-1})E_2(z) \right),$$ (5.123)

$$Y_3(z) = \frac{1}{M} \left(Y_2(z) + E_2(z) - (1 - z^{-1})E_3(z) \right).$$ (5.124)

The output of the DDSM can be expressed as

$$Y(z) = \frac{1}{M} \left(\left(1 + \frac{1}{M}(1 - z^{-1}) + \frac{1}{M^2}(1 - z^{-1})^2 \right) X(z) \right.$$

$$\left. - (1 - z^{-1})^3 \left(\frac{E_1(z)}{M^2} + \frac{E_2(z)}{M} + E_3(z) \right) + (1 - z^{-1})D(z) \right)$$ (5.125)

$$\approx \frac{1}{M} \left(X(z) - (1 - z^{-1})^3 E_3(z) \right).$$ (5.126)

Therefore,

$$E_{acc}(z) \approx -\frac{1}{M} z^{-1}(1 - z^{-1})^2 E_3(z).$$ (5.127)

Similar to the standard MASH 1-1-1,

$$Y_2(z) \approx \frac{1}{M} \left(E_1(z) - (1 - z^{-1})E_2(z) \right),$$ (5.128)

$$Y_3(z) \approx \frac{1}{M} \left(E_2(z) - (1 - z^{-1})E_3(z) \right)$$ (5.129)

due to the limited ranges of $y_1[n]$ and $y_2[n]$.

Simulation results for the MASH DDSM when the input is $X = 1$ are shown in Fig. 5.17. Similar to a standard MASH 1-1-1 DDSM, this architecture produces a quantization error signal $e_2[n]$ with a characteristic parabolic structure. In the presence of a nonlinearity, the distorted accumulated quantization error exhibits a typical wandering spur pattern.

The modified MASH DDSM proposed by Hosseini et al. has the structure shown in Fig. 5.18. This structure contains EFMs that have an extra feedback path compared to the standard EFM1. For this architecture, we have

$$Y_1(z) = \frac{1}{M} \left(\frac{1}{1 - \frac{a}{M}z^{-1}} \right) \left(X(z) - (1 - z^{-1})E_1(z) \right),$$ (5.130)

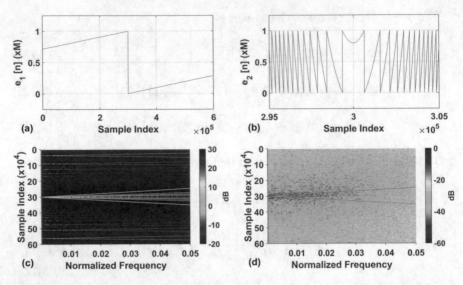

Fig. 5.17 Simulated (**a**) $e_1[n]$ waveform, (**b**) $e_2[n]$ waveform, (**c**) spectrogram of $e_2[n]/M$, and (**d**) spectrogram of $e_{acc}^{NL}[n]$ of the MASH DDSM from [38]. Input $X = 1$, modulus $M = 2^{20}$, and PWL nonlinearity of 8% mismatch is applied

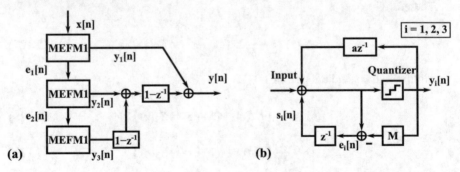

Fig. 5.18 Block diagrams of (**a**) the modified MASH DDSM from [39], (**b**) a EFM stage in the MASH DDSM

$$Y_2(z) = \frac{1}{M} \left(\frac{1}{1 - \frac{a}{M}z^{-1}} \right) \left(E_1(z) - (1 - z^{-1})E_2(z) \right), \tag{5.131}$$

$$Y_3(z) = \frac{1}{M} \left(\frac{1}{1 - \frac{a}{M}z^{-1}} \right) \left(E_2(z) - (1 - z^{-1})E_3(z) \right). \tag{5.132}$$

The combination of three outputs gives

$$Y(z) = \frac{1}{M} \frac{1}{1 - \frac{a}{M}z^{-1}} \left(X(z) - (1 - z^{-1})E_3(z) \right) \tag{5.133}$$

$$\approx \frac{1}{M} \left(X(z) - (1 - z^{-1})E_3(z) \right). \tag{5.134}$$

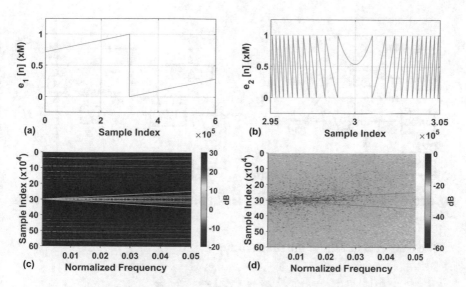

Fig. 5.19 Simulated (**a**) $e_1[n]$ waveform, (**b**) $e_2[n]$ waveform, (**c**) spectrogram of $e_2[n]/M$, and (**d**) spectrogram of $e_{acc}^{NL}[n]$ of the MASH DDSM from [39]. Input $X = 1$, modulus $M = 2^{20}$, and PWL nonlinearity of 8% mismatch is applied

The approximation is based on the fact that $a/M \approx 0$.
Thus,

$$E_{acc}(z) \approx -\frac{1}{M}z^{-1}(1 - z^{-1})^2 E_3(z). \tag{5.135}$$

Under the assumption that $a/M \approx 0$, the second terms in the output of the stages in (5.130) to (5.132) are approximately unity, leading to similar outputs to the standard MASH 1-1-1.

The simulation results for this MASH DDSM when the input $X = 1$ are shown in Fig. 5.19. Again, similar patterns in the waveforms of the internal states can be observed. The distorted accumulated quantization error shows a wandering spur pattern, as expected.

5.6.3 Other MASH DDSMs

Due to the similar output expressions, the DDSMs analyzed in the previous subsection can be seen as alternatives to a standard MASH 1-1-1. MASH DDSM with stages of unconventional transfer functions may also manifest wandering spurs. The MASH DDSM proposed by Fitzgibbon *et al.* is an example [40]. The MASH structure is shown in Fig. 5.20. For a divider controller design, we consider quantizers with a gain of $1/M$.

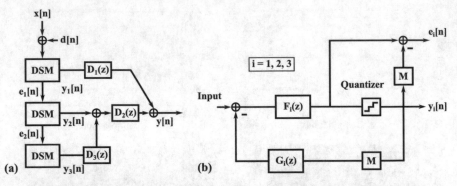

Fig. 5.20 Block diagrams of (**a**) the modified MASH DDSM from [40], (**b**) a EFM stage in the MASH DDSM

The transfer functions of the structure are

$$F_1(z) = \frac{1}{1 - z^{-1}}, \quad F_2(z) = F_3(z) = \frac{1}{1 - z^{-2}},$$

$$G_1(z) = z^{-1}, \quad G_2(z) = G_3(z) = z^{-2},$$

$$D_1(z) = 1, \quad D_2(z) = 1 - z^{-1}, \quad D_3(z) = 1 - z^{-2}. \tag{5.136}$$

Denoting the input $X_i(z)$ and the quantization error as $E_i(z)$, we have a transfer function for a stage

$$Y_i(z) = \frac{1}{M} \left(X_i(z) \frac{F_i(z)}{1 + F_i(z)G_i(z)} - \frac{1}{1 + F_i(z)G_i(z)} E_i(z) \right). \tag{5.137}$$

Substituting transfer functions in (5.136) gives

$$Y_1(z) = \frac{1}{M} \left((X(z) + D(z)) - (1 - z^{-1})E_1 \right), \tag{5.138}$$

$$Y_2(z) = \frac{1}{M} \left(E_1(z) - (1 - z^{-2})E_2 \right), \tag{5.139}$$

$$Y_3(z) = \frac{1}{M} \left(E_2(z) - (1 - z^{-2})E_3 \right). \tag{5.140}$$

First-order shaped LSB dither $D(z)$ is injected at the input of the MASH. Note that the first stage has a similar transfer function to an EFM1. This indicates that the quantization error signal is the result of the accumulation of the input and the LSB dither.

The quantization error of the second stage should be investigated next. At this stage,

$$e_2[n] - e_2[n - 2] = e_1[n] - M y_2[n]. \tag{5.141}$$

The expression for $e_2[n]$ can be found by summing equations with different indices. The result of the summation depends on the parity of n. When n is even,

$$e_2[n] = \left(e_2[-2] + \sum_{\substack{m=0, \\ m \text{ is even}}}^{n} e_1[m] \right) \bmod M. \tag{5.142}$$

When n is odd,

$$e_2[n] = \left(e_2[-1] + \sum_{\substack{m=0, \\ m \text{ is odd}}}^{n} e_1[m] \right) \bmod M. \tag{5.143}$$

These results suggest that $e_2[n]$ is related to the accumulation of $e_1[n]$ and this should lead to a parabolic structure. However, every other value in $e_1[n]$ is accumulated in this case. This is different from the case of the standard MASH 1-1-1. The wander rate of the spurs for a small input can therefore be expected to be half that in the case of a MASH 1-1-1.

Here, the accumulated quantization error can be approximated by

$$E_{acc}(z) \approx -\frac{1}{M} z^{-1} (1 - z^{-2})^2 E_3(z). \tag{5.144}$$

Consider the third stage,

$$\frac{1}{M}(e_3[n] - e_3[n-2]) = \frac{1}{M} e_2[n] - y_3[n]. \tag{5.145}$$

Therefore,

$$\frac{1}{M}(e_3[n] - 2e_3[n-2] + e_3[n-4]) =$$
$$\frac{1}{M}(e_2[n] - e_2[n-2]) - (y_3[n] - y_3[n-2]). \tag{5.146}$$

The accumulated quantization error can be expressed as

$$e_{acc}[n] \approx \frac{1}{M}(e_2[n-1] - e_2[n-3]) - (y_3[n-1] - y_3[n-3]). \tag{5.147}$$

When the accumulated quantization error is distorted by a nonlinearity, similar to the analysis for a MASH 1-1-1 DDSM, the underlying pattern in the $e_2[n]$ spectrogram may show up, which leads to visible wandering spurs.

The simulation results for this MASH DDSM are presented in Fig. 5.21. In the input $X = 1$ case, as analyzed, the wandering spur exhibits a wander rate that is

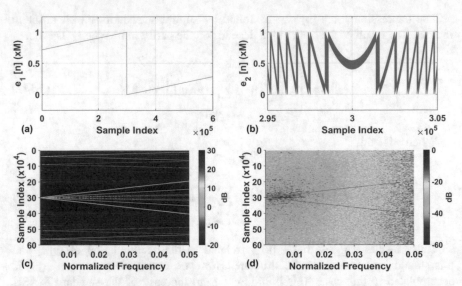

Fig. 5.21 Simulated (**a**) $e_1[n]$ waveform, (**b**) $e_2[n]$ waveform, (**c**) spectrogram of $e_2[n]/M$, and (**d**) spectrogram of $e_{acc}^{NL}[n]$ of the MASH DDSM from [40]. Input $X = 1$, modulus $M = 2^{20}$, and PWL nonlinearity of 8% mismatch is applied

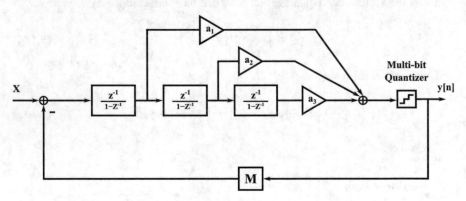

Fig. 5.22 Block diagrams the simulated CIFF DDSM. The coefficients $a_1 = 2.2$, $a_2 = 1.92$, and $a_3 = 0.72$ are used in the simulation [41]

half of that in the case of a MASH 1-1-1 DDSM and its alternatives shown in the previous subsection.

5.6.4   A Cascade-of-Integrators with Feedforward (CIFF) DDSM

In this subsection, an example of wandering spurs arising from a CIFF DDSM is given. The DDSM has the structure shown in Fig. 5.22. The output is generated from

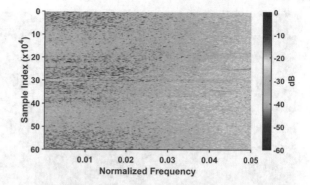

Fig. 5.23 Simulated spectrogram of $e_{acc}^{NL}[n]$ of the MASH DDSM from [40]. Input $X = 1$, modulus $M = 2^{20}$, and PWL nonlinearity of 8% mismatch is applied

the quantization of the weighted sum of the outputs of the cascaded integrators. When the input is a constant, the successive accumulation of it may lead to wandering spurs in the distorted accumulated quantization error, since all three integrator outputs contribute to the output. The simulation result for a CIFF DDSM with the input $X = 1$ is shown in Fig. 5.23. The distorted accumulated quantization error shows a typical wandering spur pattern.

5.7 Measurement Results

In this section, measurements from two commercial monolithic synthesizers are presented. The results are compared with the theoretical predictions.

5.7.1 Case I and Case II

The theoretical and simulated predictions of wandering spur patterns in Case I and Case II were verified using an Analog Devices ADF4159 fractional-N frequency synthesizer which has a MASH 1-1-1 divider controller [35]. The modulus of the MASH 1-1-1 in the ADF4159 is $M = 2^{26}$. The datasheet specifies that the MASH 1-1-1 DDSM in the ADF4159 has a nominal modulus of 2^{25}, i.e., a 25-bit MASH DDSM. Internally, the MASH DDSM has 26 bits, of which the LSB is set to one in all cases by default; the user only has access to the upper 25 bits. Therefore, the *actual* 26-bit input $X_{internal}$ is given by

$$X_{internal} = 2X_{user} + 1 \tag{5.148}$$

and $M = 2^{26}$. The inputs indicated in the measurements are the values of the internal input $X_{internal}$.

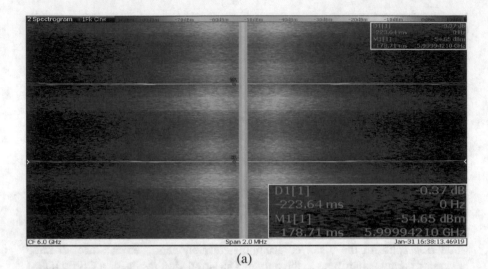

(a)

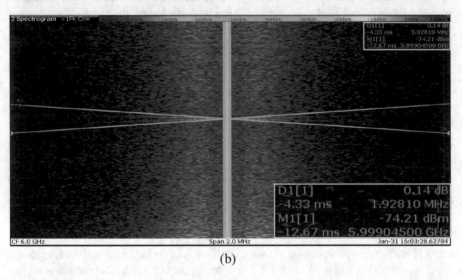

(b)

Fig. 5.24 Measured output phase noise spectrogram and a zoom of a single wandering spur event showing the X-shaped wandering spur pattern in an ADF4159 monolithic frequency synthesizer [35]. The input frequency is $f_{ref} = f_{PFD} = 100$ MHz, the MASH 1-1-1 modulus is $M = 2^{26}$, and the input is $X = 3$. The inset on the bottom right shows a zoom of the marker information

The measured wandering spur patterns when the MASH 1-1-1 input is $X = 3$ (Case I) and $X = 16777217 = M/4 + 1$ (Case II) are shown in Figs. 5.24 and 5.25, respectively. The measured fundamental wander rates and periods with the corresponding theoretical predictions are shown in Table 5.3. The measurements indicate that the proposed equations in Table 5.1 correctly predict the fundamental wander rates and the periods of the wandering spur events.

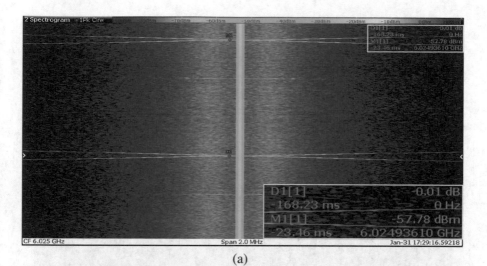

(a)

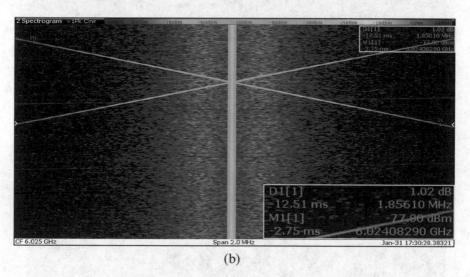

(b)

Fig. 5.25 Measured output phase noise spectrogram and a zoom of a single wandering spur event showing the X-shaped wandering spur pattern in an ADF4159 monolithic frequency synthesizer [35]. The input frequency is $f_{ref} = f_{PFD} = 100$ MHz, the MASH 1-1-1 modulus is $M = 2^{26}$, and the input is $X = 16777217$. The inset on the bottom right shows a zoom of the marker information

Table 5.3 Summary of wander rate (fundamental) and period

Input (X)	k	D	X'	Δf (MHz)	Δt (ms)	Predicted wander rate (MHz/s)	Measured wander rate (MHz/s)	Predicted period (ms)	Measured period (ms)
3	0	1	3	1.92810	4.33	447.03	445	223.71	224
16777217	1	4	1	1.85610	12.51	149.01	148	167.78	168

5.7.2   Case III

Output spectrograms were also measured from a second commercial monolithic fractional-N frequency synthesizer, namely the ADF4356 from Analog Devices [3]. The main MASH 1-1-1 DDSM in the synthesizer is 24-bit, i.e., $M = 2^{24}$, and the reference frequency of the synthesizer is set to 122.88 MHz. No dither was applied to the MASH DDSM, i.e. $d[n] = 0$.

Two spectrograms when the MASH 1-1-1 input is $X = M/2$ are shown in Fig. 5.26; the initial conditions are $s_1[0] = 1$ and $s_1[0] = \lfloor M/3 + 1 \rfloor$,

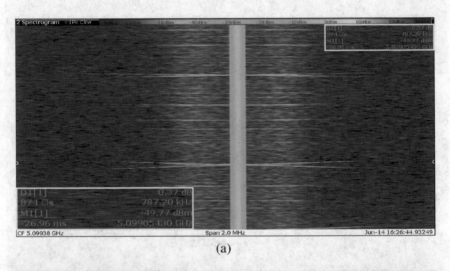

(a)

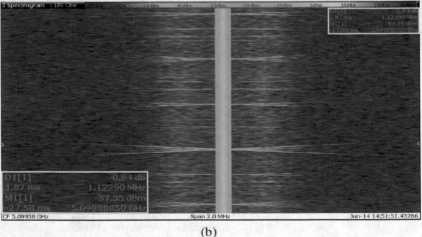

(b)

Fig. 5.26 Measured output phase noise spectrograms showing the wandering spur patterns in an ADF4356 monolithic frequency synthesizer. The input frequency is $f_{ref} = 122.88$ MHz, the MASH 1-1-1 modulus is $M = 2^{24}$ and the input is $X = M/2$. (a) $s_1[0] = 1$ and (b) $s_1[0] = \lfloor M/3 + 1 \rfloor$. The inset on the bottom left shows a zoom of the marker information

Table 5.4 Summary of wander rate (fundamental) and period

Input (X)	k	D	X'	$s_1[0]$	k'	D_{s_1}	$s_1'[0]$	Δf (MHz)	Δt (ms)	Predicted wander rate (MHz/s)	Measured wander rate (MHz/s)
8388608	1	2	0	1	0	1	1	0.7872	0.874	900	901
8388608	1	2	0	5592406	1	3	2/3	1.1229	1.87	600	600

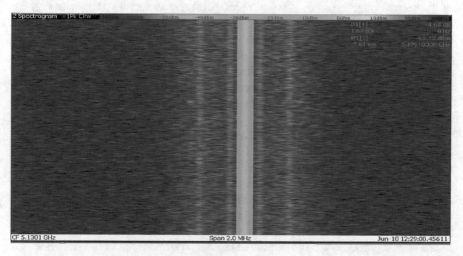

Fig. 5.27 Measured spectrogram of the output phase noise spectrum of ADF4356 with input $X = 3M/4$, $s_1[0] = 0$, $M = 2^{24}$ and a PFD update frequency $f_{PFD} = 122.88$ MHz. No dither is present. No wandering spur is observed in this case, as predicted

respectively. The predicted and measured fundamental wander rates are compared in Table 5.4.

Figure 5.27 shows the spectrogram when the input is $X = 3M/4$ and no dither is applied. No wandering spur pattern is exhibited in this case, as predicted in Sect. 5.5.3.

In all examples, the measured results match the theoretical predictions.

5.8 Summary

Wandering spur patterns in a MASH 1-1-1 DDSM-based fractional-N frequency synthesizer can be categorized into three qualitatively different types based on the input to the MASH 1-1-1.

In Case I, the accumulation of a small input directly causes a linear pattern in $e_1[n]$. This causes a parabolic structure in the quantization error of the second EFM1. The spectrogram of the quantization error from the second EFM1 contains the underlying pattern of wandering spurs seen in the output phase noise contribution.

The wandering spur events that occur at the fundamental wander rate, i.e., the slowest wander rate, are typically visible in the output phase noise contribution and these events are periodic. The fundamental wander rate and period of the events at the fundamental wander rate can be predicted.

In Case II, the residue in the accumulation of a large input causes the most apparent linear pattern in the first stage quantization error. The quantization error of the second stage contains more complex parabolic structures. Depending on the input value, different types of underlying wandering spur patterns can be observed in the spectrogram of the second stage EFM1 quantization error. The most significant events in the output phase noise spectrum occur at the fundamental wander rate and they appear with a constant period. Although different types of underlying patterns exist, the period of the most visible events can be estimated with a single equation. Thus, in Case II, the fundamental wander rate and the period of significant events at this wander rate can be predicted.

In Case III, the quantization error of the first stage EFM1 has a short period. Therefore, the initial condition of the first stage EFM1 and the dither injected into the second EFM1 stage determine the linear pattern in the internal state of the second EFM1. This leads to a parabolic pattern with components of linearly-varying frequency in the quantization error of the *third* EFM1 stage. The latter contributes to the quantization error in a direct way, causing more complex wandering spur patterns. Even with no nonlinearity present, the contribution of the MASH 1-1-1 to the output phase noise will exhibit wandering spur patterns in the presence of a non-zero initial condition and/or first-order LSB dither.

The wandering spur phenomenon results from a double accumulation in the divider controller and is therefore expected to arise in DDSMs of order at least two. The phenomenon can also be demonstrated in several different DDSMs. Neither of the conventional dither strategies for mitigating fixed spurs, namely adding zeroth-order or first-order LSB dither, can eliminate wandering spurs.

Measured spectrograms from commercial monolithic fractional-N frequency synthesizers confirm the predictions based on the theoretical analysis presented.

Chapter 6
MASH-Based Divider Controllers for Mitigation of Wandering Spurs

A DDSM divider controller that has an order greater than one may cause wandering spurs in fractional-N frequency synthesizers. In this chapter, several MASH-based divider controllers for the mitigation of wandering spurs, that both require and do not require additional dither, are presented and analyzed.

6.1 Root Cause of Wandering Spurs

In Chap. 5, wandering spurs were categorized into three cases based on the input to the MASH 1-1-1. First consider Case I and Case II where the underlying pattern of wandering spurs is present in the $e_2[n]$ spectrogram.

Figure 6.1 shows the feedforward model used to estimate the phase noise contribution of the divider controller, where the phase noise contribution (S_{DDSM}) is related to the distorted accumulated quantization error e_{acc}^{NL} of the divider controller.

Assume that the memoryless nonlinearity $NL(\cdot)$ in Fig. 6.1 can be approximated by a polynomial function, and consider a representative term $(e_{acc}[n])^p$:

$$(e_{acc}[n])^p \approx \left(-\frac{1}{M} \nabla^2 e_3[n-1] \right)^p \tag{6.1}$$

$$= \sum_{k=0}^{p} \binom{p}{k} (\nabla y_3[n-1])^k \left(-\frac{1}{M} \nabla e_2[n-1] \right)^{p-k} \tag{6.2}$$

$$= -\frac{1}{M} \nabla^2 e_3[n-1] + \frac{1}{M} \nabla e_2[n-1] + y_{3,d,cp}[n-1]$$

$$+ \sum_{k=0}^{p-1} \binom{p}{k} (\nabla y_3[n-1])^k \left(-\frac{1}{M} \nabla e_2[n-1] \right)^{p-k}, \tag{6.3}$$

© The Author(s), under exclusive license to Springer Nature Switzerland AG 2022
D. Mai, M. P. Kennedy, *Wandering Spurs in MASH-based Fractional-N Frequency Synthesizers*, Analog Circuits and Signal Processing,
https://doi.org/10.1007/978-3-030-91285-7_6

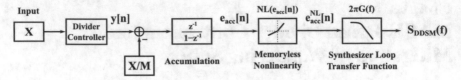

Fig. 6.1 Feedforward model with the presence of a nonlinearity

where

$$y_{3,d,cp}[n] = \begin{cases} (\nabla y_3[n])^2 - \nabla y_3[n] & \text{if } p \text{ is even,} \\ 0 & \text{if } p \text{ is odd.} \end{cases} \quad (6.4)$$

Different from the linear case where $p = 1$, apart from the $\nabla^2 e_3$ term, a number of terms associated with $e_2[n]$ are generated by the interaction with a nonlinearity, leading to the presence of wandering spurs which follow the underlying pattern in the $e_2[n]$ spectrogram [13, 15]. Notice that the first term in (6.3) is a filtered version of $e_3[n]$, i.e. it does not contain a $\nabla e_2[n]$-related term. This suggests that a wandering spur-free noise component exists in the distorted accumulated quantization error of a MASH 1-1-1. The slower-moving wandering spurs have higher amplitudes in the underlying pattern; therefore, they remain significant in the nonlinearity-induced noise. Simulations and measurements confirm that the pattern of the wandering spurs with the highest amplitudes at the fundamental wander rate appear in the output phase noise of the synthesizer [15].

Unlike Case I and Case II, the third case is characterized by a linear pattern developed in the *second* EFM1 quantization error signal $e_2[n]$ and this pattern leads to wandering spurs that show up directly in $e_3[n]$ [42]. This results in a more complicated pattern in the output phase noise spectrum. In this case, the spectrum of $e_{acc}[n]$ will exhibit the wandering spur pattern, even if the synthesizer were completely linear.

In summary, in Case I and Case II, the pattern contained in the quantization error of the second stage $e_2[n]$ is manifest in the nonlinear contribution from the divider controller to the output phase noise. In Case III, the linear pattern in $e_2[n]$ leads to a third stage quantization error that contains the wandering spur pattern. This causes the appearance of a complicated wandering spur pattern at the synthesizer output. The pattern appears even if the synthesizer itself were linear.

6.2 Dither-Based Solutions to Wandering Spurs

To reduce the wandering spur-related component when nonlinearity is present, additional dither could be considered. In a MASH 1-1-1, a 1-bit uniform dither denoted $d_1[n]$ in Fig. 6.2a is usually applied during its operation to mitigate short

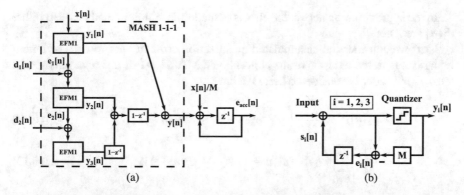

Fig. 6.2 Block diagrams of (**a**) a MASH 1-1-1 (shown in the dashed box) and (**b**) a first-order error feedback modulator (EFM1). For a conventional LSB-dithered MASH 1-1-1, the dither $d_1[n]$ is a uniformly distributed 1-bit sequence and $d_2[n]$ is absent

limit cycles [43]. An additional strong uniformly distributed dither signal $d_2[n]$ can be added at the input to the third stage EFM1, as shown in Fig. 6.2a [44]. The output of the third EFM1 in the MASH 1-1-1 can then be written as

$$y_3[n] = \frac{1}{M}\left(e_2[n] - \nabla e_3[n] + d_2[n]\right), \tag{6.5}$$

$$Y_3(z) = \frac{1}{M}\left(E_2(z) - (1 - z^{-1})E_3(z) + D_2(z)\right) \tag{6.6}$$

in the time and frequency domains, respectively.

The output of the MASH 1-1 1 is

$$y[n] = y_1[n] + \nabla y_2[n] + \nabla^2 y_3[n], \tag{6.7}$$

where ∇^k denotes the kth backward difference. Specifically,

$$\nabla y_2[n] = y_2[n] - y_2[n-1], \tag{6.8}$$

$$\nabla^2 y_3[n] = \nabla y_3[n] - \nabla y_3[n-1]$$
$$= y_3[n] - 2y_3[n-1] + y_3[n-2]. \tag{6.9}$$

The accumulated quantization error in this case is given by

$$e_{acc}[n] = \frac{1}{M}\left(-\nabla^2 e_3[n-1] + \nabla d_2[n-1] + s_1[0] + d_1[n-1]\right) \tag{6.10}$$

$$\approx \frac{1}{M}\left(-\nabla^2 e_3[n-1] + \nabla d_2[n-1]\right). \tag{6.11}$$

In the approximation above, the effects of the initial condition and the LSB dither $d_1[n]$ are ignored.

Next we consider the accumulated quantization error in the presence of a non-linearity. As in the analysis of the conventional MASH 1-1-1, a simple polynomial term of pth order is considered here. We have

$$(e_{acc}[n])^p \approx \frac{1}{M^p} \left(-\nabla^2 e_3[n-1] + \nabla d_2[n-1]\right)^p \tag{6.12}$$

$$= \frac{1}{M^p} \sum_{k=0}^{p} \binom{p}{k} (\nabla d_2[n-1])^k (-\nabla^2 e_3[n-1])^{p-k}. \tag{6.13}$$

Notice that the $(\nabla d_2[n-1])^p$ term is independent of the quantization error $e_2[n]$ from the the previous stage. We next consider the $e_3[n]$-related terms.

The EFM1 is essentially an accumulator with a modulus of M. Therefore,

$$e_3[n] = \left(s_3[0] + \sum_{k=0}^{n} q[k] \right) \mod M, \tag{6.14}$$

where

$$q[n] = (e_2[n] + d_2[n]) \mod M. \tag{6.15}$$

The equivalent input to the third EFM can be seen as the quantized sum of $e_2[n]$ and $d_2[n]$ due to the property of modulo operation, as given in (6.15).

Now consider the impact of the range of the additional dither $d_2[n]$. When the range of $d_2[n]$ is very limited, e.g., a 1-bit stream, since $e_2[n] \in [0, M-1]$, the quantization effect is trivial and $q[n] \approx e_2[n]$ [45]. For moderate ranges of $d_2[n]$, the cases where the sum of $e_2[n]$ and $d_2[n]$ is greater than M will increase; consequently, the effect of the quantization becomes significant. The input can be seen as a combination of $e_2[n]$ and a noise signal $e[n]$, i.e., $q[n] = e_2[n] + e[n]$. When the range of the dither is sufficiently large, the quantization effect of the sum of the two inputs to the third stage is dominant and the equivalent input that is related to the generation of the quantization error signal $e_3[n]$ can be seen as a random sequence $d[n]$ that takes values between 0 and $M-1$. In this case, $q[n] \approx d[n]$. Due to the negligible correlation between $q[n]$ and $e_2[n]$, the e_3-related terms in (6.13) can be expected to be dominated by noise that is free of wandering spurs.

Also, as the range of $d_2[n]$ increases, the $d_2[n]$ component in (6.13) increases, resulting in more uncorrelated noise. The property of the exponents leads to a significant increase of the noise component in the distorted accumulated quantization error when the maximum of $d_2[n]$ is increased to be greater than the modulus M.

Based on the analysis above, a requirement that the range of the dither is close to the modulus M is reasonable. Since $e_2[n] \in [0, M-1]$, a dither signal $d_2[n]$ with a range of $[0, M-1]$ can lead to the sum $(e_2[n] + d_2[n]) \mod M$ having any

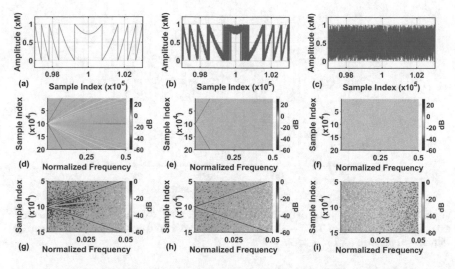

Fig. 6.3 The waveform of $q[n]$, the spectrogram of $q[n]$, and the spectrogram of $e_{acc}^{NL}[n]$. (**a**), (**d**) and (**g**): no additional dither; (**b**), (**e**) and (**h**): uniform $d_2[n]$ with range $[0, M/4 - 1]$; (**c**), (**f**) and (**i**): uniform $d_2[n]$ with range $[0, M - 1]$. Input $X = 1$ and $M = 2^{20}$ in all cases

value between zero and $M - 1$. The sum under the modulo operation will become similar to $e_2[n]$ when $d_2[n]$ with a narrower range is applied. For the convenience of implementation, it is required that the uniformly distributed random number $d_2[n]$ should have the range $[0, 2^k - 1]$, where $k = 1, 2 \ldots$. Such a dither signal $d_2[n]$ can be generated by an LFSR-based pseudorandom number generator, for example.

Figure 6.3 shows the waveform and spectrogram of $q[n]$ and the spectrogram of $e_{acc}^{NL}[n]$, i.e., $e_{acc}[n]$ distorted by the nonlinearity, of a standard MASH 1-1-1, a MASH 1-1-1 with a $d_2[n]$ range of $[0, M/4 - 1]$, and a MASH 1-1-1 with a $d_2[n]$ range of $[0, M - 1]$, where the modulus is $M = 2^{20}$.[1] A PWL model of a charge pump nonlinearity with 8% static mismatch between the up and down currents is applied, as in the following simulations, unless otherwise stated.[2] The simulations confirm that additive dither $d_2[n]$ with a range of $[0, M - 1]$ is sufficient to eliminate the wandering spur pattern.

Note that the extra dither requires a wider output $y_3[n]$ range of $[0, 2]$, yielding a range of $[-5, 6]$ for the output $y[n]$. This indicates that the bus widths of the output and the combination filter must be increased accordingly, compared to the standard EFM1-based MASH 1-1-1. The impact of the extended output range will be discussed in later sections.

---

[1] Each normalized spectrogram shown consists of modified periodogram estimates from segments of 4000 points. A Hanning window is applied and segments have an overlap of 95%.

[2] In theoretical analysis, 8% static mismatch is considered significant [5, 27]. For implementation, charge pump mismatch of about 1% to 5% can be expected for a reasonably good design [31, 46].

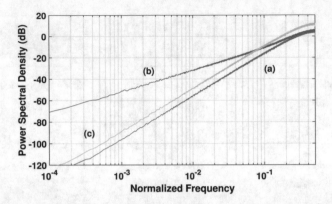

Fig. 6.4 The spectra of $e_{acc}[n]$ of (**a**) the standard MASH 1-1-1, (**b**) a MASH 1-1-1 with uniform dither $d_2[n]$ with a range of $[0, M-1]$, and (**c**) a MASH 1-1-1 with first-order shaped dither $d_2[n]$ with a $d_2'[n]$ range of $[0, 2M-1]$, $M = 2^{20}$

It should be noticed that, in the linear case, the output and accumulated quantization error in (6.11) have z-domain equivalents of

$$Y(z) = \frac{1}{M}(X - (1 - z^{-1})^3 E_3(z) + (1 - z^{-1})^2 D_2(z)), \tag{6.16}$$

and

$$E_{acc}(z) \approx \frac{1}{M}\left(-(1 - z^{-1})^2 E_3(z) + (1 - z^{-1}) D_2(z)\right), \tag{6.17}$$

respectively.

When the dither $d_2[n]$ is dominant, the first-order shaped noise dominates the phase noise contribution of the divider controller in the linear case, as shown in Fig. 6.4. The standard MASH 1-1-1 has a second-order shaped phase noise contribution in the linear case, which gives it a theoretical advantage in terms of its in-band noise performance.

In order to achieve third-order shaping of the quantization error and to lower the in-band contribution, a shaped dither signal $d_2[n]$ generated by high-pass filtering can be applied. In the frequency domain, a dither signal

$$D_2(z) = (1 - z^{-1})^k D_2'(z), \ k = 1, \ 2, \ \ldots \tag{6.18}$$

where $D_2'(z)$, or $d_2'[n]$ in the time domain, is a uniform dither source, can be applied. Choosing $k = 1$,

$$d_2[n] = \nabla d_2'[n] = d_2'[n] - d_2'[n-1]. \tag{6.19}$$

The shaped additional strong dither contributes to the quantization error signal $e_{acc}[n]$ in the presence of a nonlinearity differently and the range that is sufficient to

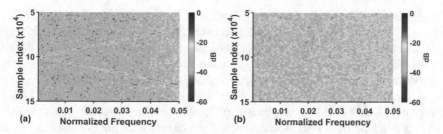

Fig. 6.5 The spectrogram of $e_{acc}^{NL}[n]$: (**a**) MASH 1-1-1, (**b**) first-order high-pass shaped $d_2[n]$ with a $d_2'[n]$ range of $[0, M-1]$, input $X = \lfloor M/3+2 \rceil = 349527$ and $M = 2^{20}$. $\lfloor \cdot \rceil$ stands for rounding to the nearest integer

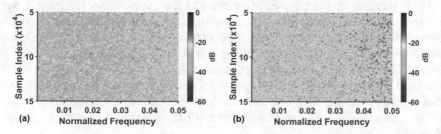

Fig. 6.6 The spectrogram of $e_{acc}^{NL}[n]$: (**a**) first-order high-passed $d_2[n]$ with a $d_2'[n]$ range of $[0, M-1]$, (**b**) first-order high-passed $d_2[n]$ with a $d_2'[n]$ range of $[0, 2M-1]$, input $X = M/2$, $s_1[0] = 1$, and $M = 2^{20}$

eliminate wandering spurs should be determined again. A dither with a $d_2'[n]$ range of $[0, M-1]$ is first considered. As shown in Fig. 6.5, simulations indicate that it is capable of wandering spur-free performance for Case I and II inputs. Figure 6.6 shows the spectrogram of $e_{acc}[n]$ for the MASH 1-1-1 with $d_2'[n]$ having ranges $[0, M-1]$ and $[0, 2M-1]$ in the presence of a PWL nonlinearity of 8% mismatch. The weak wandering spur that appears in the spectrogram in Fig. 6.6a indicates that an additional dither with a source range of $[0, M-1]$ is not sufficient to eliminate the wandering spurs in all cases. As suggested by these simulations, by increasing the range of $d_2'[n]$ to $[0, 2M-1]$, wandering spur-free performance can be achieved for all inputs. Note that, with this dither solution, the range of the output $y[n]$ is extended to $[-11, 12]$, compared with the output range of $[-3, 4]$ of a standard MASH 1-1-1.

6.3   Modified MASH DDSM Solutions to Wandering Spurs

The dither solution to the wandering spur in the MASH 1-1-1 described in Sect. 6.2 requires a full-scale or larger external source of uniformly distributed random numbers to generate the dither signal. Furthermore, as (6.17) indicates, the spectral performance of the dither $d_2[n]$ can impact or even dominate the divider controller noise. Therefore, this adds complexity to the originally simple design of a MASH 1-1-1 since a pseudorandom number generator of sufficient performance, which is usually LFSR-based, is required. A more concise solution that involves only modification of one EFM1 stage is proposed and analyzed in this section.

In Sect. 6.2, it has been shown that an additional uniform dither signal with a range of at least $[0, M - 1]$ can be applied to eliminate the wandering spur pattern. The analysis of the EFM1 structure usually assumes that the quantization error $e_i[n]$ is white noise. However, due to the simple accumulation of the input, the EFM1 quantization error is correlated with its input, as analyzed in Sect. 6.1, and this causes wandering spurs in the presence of nonlinearity.

It is possible to alter the EFM1 such that the quantization error has lower correlation with its input, which is the quantization error from the previous EFM1. Since the additional dither that is required to eliminate wandering spurs is injected at the final stage, we consider modifying the *third* EFM1 of the standard MASH 1-1-1. This is sensible since the output equations (4.1), (4.2), and (4.3), as well as the combination equation for the output (6.7), suggest that modifications of the first and second EFM1 feedback transfer functions would require corresponding changes to the error cancellation network transfer functions, while modifying the third stage does not.

6.3.1 Second-Order Solutions

Assume that $e_3[n]$ is uniformly distributed within $[0, M - 1]$ and has negligible correlation with $e_2[n]$.[3] Instead of adding an external dither $d_2[n]$, suitable equivalent signals can be generated by passing $E_3(z)$ through a dither transfer function $DT(z)$, i.e.,

$$D_2(z) = DT(z)E_3(z). \tag{6.20}$$

An equivalent dither signal with a range of at least $[0, M - 1]$ can be produced by:

1. Increasing the coefficient of z^{-1} in the feedback,
2. Adding a further delayed quantization error,
3. Adding a combination of delayed quantization errors.

---

[3] This assumption holds for the modified MASH structures presented and can be verified by simulations.

6.3.1.1   Increasing the Coefficient of z^{-1} in the Feedback

First consider simply increasing the coefficient in the feedback path:

$$DT(z) = cz^{-1}, \ c = 1, \ 2, \ \ldots . \tag{6.21}$$

giving

$$D_2(z) = cz^{-1}E_3(z) \tag{6.22}$$

The $c = 1$ case gives

$$D_2(z) = z^{-1}E_3(z). \tag{6.23}$$

Figure 6.7 shows the structure of the modified EFM1 and the output of the stage is

$$Y_3(z) = \frac{1}{M}\left(E_2(z) - (1 - z^{-1} - DT(z))E_3(z)\right) \tag{6.24}$$

$$= \frac{1}{M}\left(E_2(z) - (1 - FT(z))E_3(z)\right), \tag{6.25}$$

where $FT(z) = z^{-1} + DT(z)$ is the feedback transfer function.
This equivalent uniform dither gives a third stage output of

$$Y_3(z) = \frac{1}{M}(E_2(z) - (1 - 2z^{-1})E_3(z)). \tag{6.26}$$

The equivalent dither in this case satisfies the assumption of an independent uniform dither, giving a significantly attenuated, but still visible, pattern in the spectrogram when compared with the standard MASH 1-1-1, as shown in Fig. 6.8b.

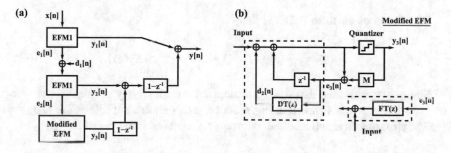

Fig. 6.7 The block diagram illustrating (**a**) the modified MASH-based divider controller and (**b**) the modified EFM

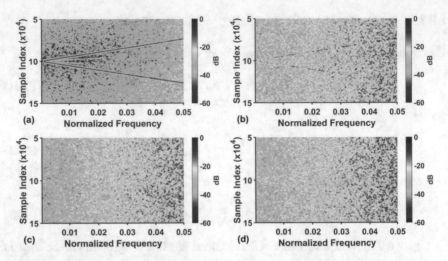

Fig. 6.8 The spectrograms of $e_{acc}^{NL}[n]$ of MASH-based divider controllers. (**a**) Standard MASH 1-1-1, (**b**) modified MASH 1-1-1 with $D_2(z) = z^{-1}E_3(z)$, (**c**) modified MASH 1-1-1 with $D_2(z) = z^{-2}E_3(z)$, (**d**) modified MASH 1-1-1 with $D_2(z) = (-z^{-1} + 2z^{-2})E_3(z)$; input $X = 2$ and $M = 2^{20}$

6.3.1.2 Adding a Further Delayed Quantization Error

Another possible equivalent dither is a further-delayed version of $e_3[n]$:

$$D_2(z) = z^{-k}E_3(z), \ k = 2, \ 3, \ \dots . \tag{6.27}$$

For example, the quantization error delayed by two samples can be considered as equivalent dither with an identical range:

$$D_2(z) = z^{-2}E_3(z). \tag{6.28}$$

The output of the third EFM1 is then

$$Y_3(z) = \frac{1}{M}(E_2(z) - (1 - z^{-1} - z^{-2})E_3(z)). \tag{6.29}$$

Simulation indicates that this equivalent dither effectively eliminates the wandering spur pattern, as shown in Fig. 6.8c. Negative equivalent dither, i.e. $D_2(z) = -z^{-k}E_3(z)$, has an identical effect but requires a subtraction operation.

6.3.1.3 Adding a Combination of Delayed Quantization Errors

A combination of delayed quantization errors,

$$D_2(z) = (\alpha z^{-k} + \beta z^{-l}) E_3(z), \tag{6.30}$$

where α, β, k, and l are integers with $l > k \geq 1$ and $|\alpha + \beta| = 1$, can be applied as the equivalent dither.

Noticing that the EFM1 already contains a delayed quantization error signal $e_3[n-1]$, one could consider

$$D_2(z) = (-z^{-1} + 2z^{-k}) E_3(z), \quad k = 2, \ 3, \ \ldots. \tag{6.31}$$

Choosing $k = 2$, we have

$$D_2(z) = (-z^{-1} + 2z^{-2}) E_3(z), \tag{6.32}$$

which can be applied as an effective equivalent and this gives

$$Y_3(z) = \frac{1}{M}(E_2(z) + (1 - 2z^{-2}) E_3(z)). \tag{6.33}$$

As shown in Fig. 6.8d, this equivalent dither is effective.

To summarize, three types of equivalent $d_2[n]$ dither are shown to be capable of mitigating the wandering spurs. The transfer functions $DT(z)$ implementing different equivalent dithers are not equally efficient in eliminating the wandering spur pattern. The equivalent dither $d_2[n]$ produced by simply increasing the coefficient in the feedback path of EFM1 is the least effective. A feedback of $FT(z) = 2z^{-k}$ can be adopted as an acceptable design. Of the three examples considered here, the equivalent dither $D_2(z) = z^{-2} E_3(z)$ achieves the best performance.

6.3.2 Third-Order Solutions

In Sect. 6.2, to lower the in-band noise caused by uniform dither, a first-order shaped dither with a range of $[0, 2M - 1]$ was applied. One could modify the feedback transfer function of the third EFM1 in the MASH 1-1-1 to achieve similar spectral performance.

Higher order equivalent dither designs can be constructed based on the uniform equivalents given in previous designs. Following (6.19), we consider the following equivalents for $d_2'[n]$ in the generation of shaped dither:

$$D_{2,1}'(z) = 2z^{-1}E_3(z), \tag{6.34}$$

$$D_{2,2}'(z) = 2z^{-2}E_3(z), \tag{6.35}$$

$$D_{2,3}'(z) = 2(-z^{-1} + 2z^{-2})E_3(z). \tag{6.36}$$

Following the method of generating shaped noise in Sect. 6.2,

$$D_2(z) = (1 - z^{-1})^k D_{2,i}'(z), \; k = 1, \; 2, \; \ldots \tag{6.37}$$

can be added in the feedback of the third EFM1.

Consider $k = 1$. First-order shaping generates both positive and negative values, leading to possible equivalent shaped dithers of

$$D_2(z) = (1 - z^{-1})D_{2,i}'(z) \tag{6.38}$$

and

$$D_2(z) = -(1 - z^{-1})D_{2,i}'(z), \tag{6.39}$$

where $i = 1, 2, 3$. Various designs are possible and they lead to different noise profiles and spreads of the output of the modified MASH. Here we consider two representative designs.

The first design has the equivalent dither

$$D_2(z) = -2z^{-1}(1 - z^{-1})E_3(z). \tag{6.40}$$

The output for the EFM in this case is

$$Y_3(z) = \frac{1}{M}(E_2(z) - (1 + z^{-1} - 2z^{-2})E_3(z)) \tag{6.41}$$

$$= \frac{1}{M}(E_2(z) - (1 + 2z^{-1})(1 - z^{-1})E_3(z)). \tag{6.42}$$

The quantization errors $e_2[n]$ and $e_3[n]$ both have a range of $[0, M - 1]$ and the feedback transfer function is

$$FT(z) = -z^{-1} + 2z^{-2}. \tag{6.43}$$

The range of the output $y[n]$ in this design is $[-7, 8]$. This is a narrower range compared to the MASH 1-1-1 with extra high-pass shaped dither $d_2[n]$ that was

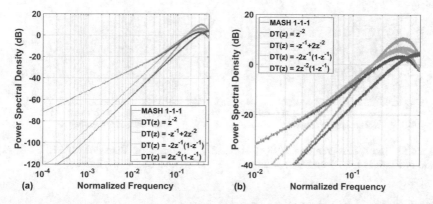

Fig. 6.9 Simulated linear $e_{acc}[n]$ spectra of representative second-order (yellow and red) and third-order (cyan and green) solutions: (**a**) full spectra, (**b**) zoomed spectra around the peaks. The spectrum of the standard MASH 1-1-1 is shown in blue

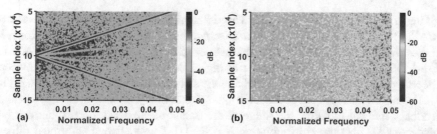

Fig. 6.10 The spectrograms of $e_{acc}^{NL}[n]$ of MASH structures: (**a**) Standard MASH 1-1-1, (**b**) modified MASH 1-1-1 with $D_2(z) = -2z^{-1}(1 - z^{-1})E_3(z)$, input $X = 1$ and $M = 2^{20}$

described in Sect. 6.2. As shown in Fig. 6.9a, the design gives a second-order shaped accumulated quantization error which is higher than the profile of the standard MASH 1-1-1. A comparison of the spectrograms of the standard MASH 1-1-1 and this modified MASH is shown in Fig. 6.10. The standard MASH 1-1-1 suffers from wandering spurs while the modified structure does not.

Another representative equivalent shaped dither is

$$D_2(z) = 2z^{-2}(1 - z^{-1})E_3(z), \tag{6.44}$$

which leads to an output of

$$Y_3(z) = \frac{1}{M}(E_2(z) - (1 - z^{-1} - 2z^{-2} + 2z^{-3})E_3(z)) \tag{6.45}$$

$$= \frac{1}{M}(E_2(z) - (1 - 2z^{-2})(1 - z^{-1})E_3(z)). \tag{6.46}$$

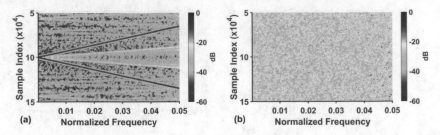

Fig. 6.11 The spectrograms of $e_{acc}^{NL}[n]$ of MASH structures: (**a**) Standard MASH 1-1-1, (**b**) modified MASH 1-1-1 with $D_2(z) = 2z^{-2}(1 - z^{-1})E_3(z)$, input $X = M/2$ and $M = 2^{20}$

The feedback transfer function for this design is

$$FT(z) = z^{-1} + 2z^{-2} - 2z^{-3}, \tag{6.47}$$

giving a $y[n]$ range of $[-11, 12]$. This is significantly larger than the standard MASH 1-1-1, the MASH 1-1-1 with uniform dither, and its equivalents. Figure 6.9b shows that the design has similar theoretical in-band performance to the MASH 1-1-1 DDSM, but with higher noise at high frequencies. A comparison between spectrograms of this modified MASH and the standard MASH is shown in Fig. 6.11.

These two designs show that a trade-off exists: by choosing an appropriate feedback transfer function, the range of the output of the divider controller can be made smaller than the corresponding MASH 1-1-1 with shaped additional dither but the noise is higher than the standard MASH 1-1-1. Other architectures based on different choices of $D_2'(z)$ have wider output ranges but the in-band phase noise performance is closer to the ideal second-order shaped noise.

6.3.3   Summary

Table 6.1 summarizes the second-order and third-order modified MASH structures described in this section. The linear accumulated quantization error ($e_{acc}[n]$) spectra of these designs are shown in Fig. 6.9. It should be noted that the proposed wandering spur solutions all have wider ranges of output $y[n]$ than the standard MASH 1-1-1.[4]

---

[4] In the literature, divider controllers with extended output ranges compared to standard MASH DDSMs, e.g., the architectures in [26] and [47], have been implemented. In such cases, the divider division ratio range required is increased and the overall design of the synthesizer should be adjusted to avoid extremely small division ratios in fractional-N mode. Also, the nonlinearity of the PFD and charge pump should be minimized for optimum spectral performance.

Table 6.1 Table of the transfer functions of representative MASH DDSMs

Dither transfer function $DT(z)$	Feedback transfer function $FT(z)$	Third EFM noise transfer function	Range of output $y[n]$	Comment
0	z^{-1}	$1 - z^{-1}$	$[-3, 4]$	Standard MASH 1-1-1
z^{-1}	$2z^{-1}$	$1 - 2z^{-1}$	$[-5, 6]$	Shows wandering spur pattern
z^{-2}	$z^{-1} + z^{-2}$	$1 - z^{-1} - z^{-2}$	$[-5, 6]$	First second-order solution
$-z^{-1} + 2z^{-2}$	$2z^{-2}$	$1 - 2z^{-2}$	$[-5, 6]$	Second second-order solution
$-2z^{-1}(1 - z^{-1})$	$-z^{-1} + 2z^{-2}$	$(1 + 2z^{-1})(1 - z^{-1})$	$[-7, 8]$	First third-order solution
$2z^{-2}(1 - z^{-1})$	$z^{-1} + 2z^{-2} - 2z^{-3}$	$(1 - 2z^{-2})(1 - z^{-1})$	$[-11, 12]$	Second third-order solution

6.4   Evaluation of Spectral Performance of the Representative Designs

6.4.1   Envelope of Short-Term Spectrum

The proposed solutions can reduce the variation in output phase noise caused by wandering spurs. This can be observed in the spectral envelope defined by the maximum values of the short-term spectra, as shown in Fig. 6.12. Simulation of the synthesizer with the parameters in [26] shows that the spectral envelope of the output phase noise of the synthesizer with the standard MASH 1-1-1 is determined by the wandering spurs at low offset frequencies; this is illustrated in Fig. 6.12a. The synthesizer based on the modified MASH with $D_2(z) = z^{-2}E_3(z)$ does not exhibit wandering spurs in simulation and therefore it can achieve a maximum reduction of 15 dB in the envelope (shown in blue in Fig. 6.12c and d) when compared with the standard MASH-based synthesizer, as shown in Fig. 6.12b and d.

6.4.2   Noise Floor in Long-Term Spectrum

Compared to the standard MASH 1-1-1, the proposed designs can lead to a rise in the in-band noise in the long-term spectrum in the presence of a nonlinearity due to their extended output ranges [5]. A comparison of $e_{acc}[n]$ spectra of representative MASH designs in the presence of a PWL nonlinearity representing 8% mismatch is

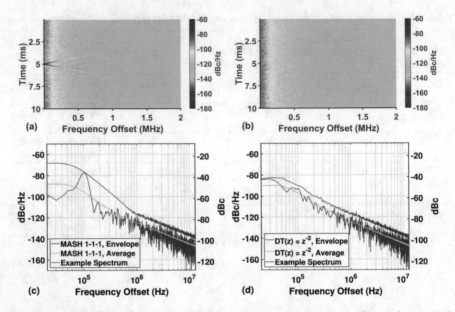

Fig. 6.12 The simulated output phase noise spectrogram of a synthesizer with (**a**) the standard MASH 1-1-1 and (**b**) the modified MASH with $D_2(z) = z^{-2}E_3(z)$. The spectral envelopes of the output phase noise of the synthesizer with the standard MASH 1-1-1 and the modified MASH with $D_2(z) = z^{-2}E_3(z)$ are shown in (**c**) and (**d**), respectively. Input $X = 1$ and modulus $M = 2^{20}$. A PWL nonlinearity with 8% mismatch is used in the simulations

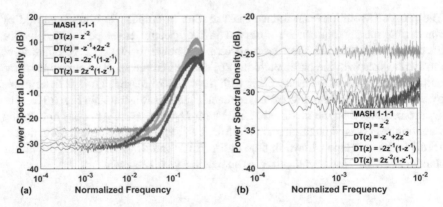

Fig. 6.13 Comparison of $e_{acc}[n]$ spectra in the presence of a PWL nonlinearity with 8% mismatch of example second-order solutions and third-order solutions: (**a**) full spectra and (**b**) zoomed spectra of the noise floors

shown in Fig. 6.13. The simulations show that the folded noise from the third-order solutions introduces a higher noise floor than the second-order designs. In the worst case, a contribution to the low-frequency noise floor that is 6 dB higher than that of the standard MASH is observed.

6.4.3   Fixed Spurs in Long-Term Spectrum

We next consider the performance of the designs in terms of fixed spurs. The fixed spurs caused by the standard MASH 1-1-1 DDSM divider are mostly related to the periodicity resulting from the accumulation of the constant input in the first stage EFM1. The dither solutions and the modified MASH solutions involve modification to the later stages. Therefore, the proposed designs would still produce fixed spurs related to the accumulation of the input.

It should be noted that the extended output range of the divider controller can lead to the change in the range of the accumulated quantization error, which may also affect the stationary spur performance [9]. These wandering spur solutions directly or indirectly introduce noise at the input of the third EFM1; this reduces the correlation between the equivalent input $q[n]$ and the quantization error $e_2[n]$. The $\nabla e_2[n]$ sequence, which contributes to the distorted accumulated quantization error as indicated by (6.3), can be written as

$$\nabla e_2[n] = e_1[n] - My_2[n], \tag{6.48}$$

i.e., $\nabla e_2[n]$ is correlated with $e_1[n]$, which is periodic in the case of a fixed input X and therefore causes spurious tones in the long-term spectrum of the divider controller noise. Reducing the $e_2[n]$-related contribution to the accumulated quantization error may also help to reduce the correlation with prior stages, resulting in lower amplitude periodic components that produce spurs.

In a scenario of 1% mismatch rather than 8%, and with the parameters used in [26], the long-term spectra of the output phase noise of synthesizers based on a standard MASH 1-1-1 and the design with $DT(z) = 2z^{-2}(1 - z^{-1})$ have similar profiles, as shown in Fig. 6.14. Note that the modified MASH exhibits a fixed spur of -67 dBc while the corresponding spur caused by the standard MASH 1-1-1 has an amplitude of -61 dBc. Thus, the MASH modification that eliminates the wandering spurs may also reduce the amplitudes of fixed spurs.

Fixed spurs can also arise from the regrowth of components related to the second-stage internal state $e_2[n]$ in the presence of nonlinearity. Such fixed tones can be observed in Case III, as shown in Fig. 6.15a. Since the dither solutions and modified MASH structures reduce the components related to $e_2[n]$ when nonlinearity is present, these fixed spurs can be eliminated along with the wandering spurs, as shown in Fig. 6.15b.

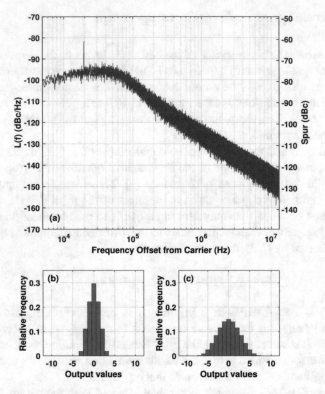

Fig. 6.14 (**a**) Output phase noise from a synthesizer with a MASH 1-1-1 (blue) and a synthesizer with the modified MASH (red) with $D_2(z) = 2z^{-2}(1 - z^{-1})E_3(z)$ and the output distribution of (**b**) MASH 1-1-1 and (**c**) modified MASH with $D_2(z) = 2z^{-2}(1 - z^{-1})E_3(z)$

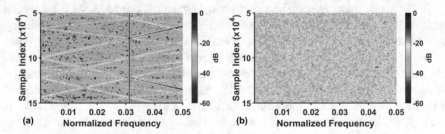

Fig. 6.15 The spectrograms of $e_{acc}^{NL}[n]$ of MASH structures: (**a**) Standard MASH 1-1-1, (**b**) modified MASH 1-1-1 with $D_2(z) = 2z^{-2}(1 - z^{-1})E_3(z)$, input $X = M/16$ and $M = 2^{20}$

6.5   Summary

Wandering spurs in a MASH 1-1-1-based conventional fractional-N frequency synthesizer originate in the DDSM divider controller. The quantization error of the second EFM1 $e_2[n]$ is the root cause of wandering spurs. In Case I and Case

II, $e_2[n]$ has the underlying wandering spur pattern in its spectrogram and this indirectly contributes to the output phase noise contribution of the MASH 1-1-1 in the presence of nonlinearity. In Case III, $e_2[n]$ contains a linear pattern that leads to the parabolic structure and therefore the wandering spur pattern of the third stage quantization error $e_3[n]$. Since $e_3[n]$ is directly related to the phase noise contribution of the MASH, the complex spectrogram pattern can be observed.

An additional uniform dither signal $d_2[n]$ injected into the MASH 1-1-1 at the input of the third EFM1 can mitigate wandering spurs. With large amplitude dither $d_2[n]$, the equivalent input to the third EFM1 becomes sufficiently uncorrelated with the quantization error signal $e_2[n]$ from the previous EFM1. The increase in the range of the uniform dither signal $d_2[n]$ also adds uncorrelated noise to the accumulated quantization error in both the linear and nonlinear cases. Therefore, applying a dither $d_2[n]$ with a sufficient range can effectively mitigate wandering spurs in the presence of a given nonlinearity. A uniformly distributed random signal $d_2[n]$ with a range of $[0, M-1]$ can eliminate the wandering spur in a MASH 1-1-1 in the presence of a PWL nonlinearity with 8% mismatch.

A high-pass filtered dither signal can be applied instead for lower low-frequency noise. A dither $d_2[n] = d_2'[n] - d_2'[n-1]$ with uniformly distributed $d_2'[n]$ having a range of $[0, 2M-1]$ can achieve the same result of wandering spur mitigation.

A more compact design can be realized by modifying the standard MASH 1-1-1 architecture. The feedback transfer function of the third EFM1 is modified in order to minimize the changes to the standard MASH 1-1-1. By assuming that the quantization error $e_3[n]$, which has a range of $[0, M-1]$, is uniformly distributed and uncorrelated with $e_2[n]$, an equivalent dither can be constructed with a transfer function $DT(z)$, i.e., $D_2(z) = DT(z)E_3(z)$. The effectiveness of the MASH structure in mitigating wandering spurs depends on the choice of transfer function.

Similar to the dither solution, the transfer function of the third EFM1 can be modified to achieve a higher-order shaped quantization error. A trade-off between the phase noise contribution and the output range of the MASH divider controller can be achieved by selecting different transfer functions.

The dither solution and its equivalents expand the output range of the MASH divider controller and reduce the effect of noise contributions from previous stages. This can lead to better performance in terms of stationary spurs but will potentially increase the in-band noise. In terms of implementation, one should note the corresponding change in the range of the multi-modulus divider division ratio and the PFD/CP input. The PFD/CP should be as linear as possible over the extended range of input for optimum performance.

In applications where DDSMs have significant output spectral contribution, for example audio applications [12], the proposed designs can be used in place of a standard DDSM to improve the spectral performance.

Chapter 7
Measurement Results for MASH-Based Wandering Spur Mitigation Solutions

In this chapter, measurement results for a demonstrator synthesizer with MASH-based wandering spur mitigation solutions are presented. The details of the implemented synthesizer and divider controller structure are presented first. Results for the MASH-based divider controllers in all three cases of wandering spurs are shown and their impacts on the output phase noise performance are compared.

7.1 Implemented Synthesizer and Divider Controller Structure

7.1.1 The Fractional-N Frequency Synthesizer

The block diagram of the synthesizer is shown in Fig. 7.1. The platform on which it is built is the latest generation of charge-pump type-II fractional-N frequency synthesizer produced by Analog Devices; it is implemented in a 180 nm SiGe BiCMOS process [48]. Apart from the charge pump current sources, the synthesizer contains a series of binary weighted current sources to provide a variable bleed current. A quad-core pseudo-differential Colpitts VCO can generate output frequencies ranging from 4 GHz to 8 GHz. Dividers and multipliers are available at the output of the synthesizer for different output ranges. No output frequency divider or multiplier is used for the measurements presented in this chapter.

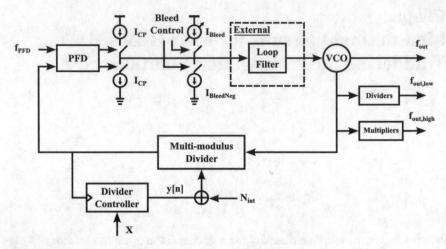

Fig. 7.1 Block diagram of the implemented fractional-N frequency synthesizer

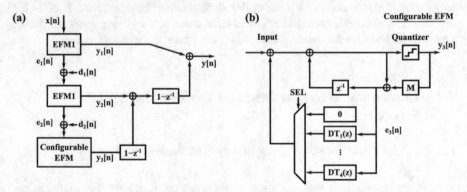

Fig. 7.2 Block diagram of the implemented MASH DDSM divider controller

7.1.2   MASH-Based Divider Controllers

The divider controller in the fractional-N frequency synthesizer demonstrator can be configured to realize different MASH-based architectures. The structure of the divider controller is shown in Fig. 7.2a. Dither $d_1[n]$ is a 1-bit LSB dither and dither $d_2[n]$ is a full-range uniform dither between 0 and $(M - 1)$ generated from an external linear feedback shift register (LFSR). A 61-bit LFSR is used to generate the dither signals. To provide the full-scale uniform dither $d_2[n]$, the output of the LFSR is scrambled using XOR operations.

The MASH has three stages. The first two stages are conventional EFM1s and the third stage is a configurable EFM. The feedback transfer function can be changed by selecting different configurations using the SEL signal, as shown schematically in Fig. 7.2b.

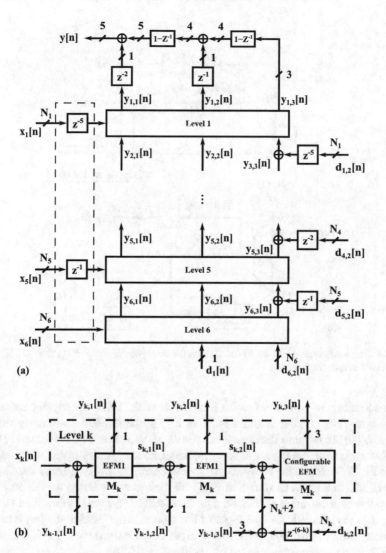

Fig. 7.3 (**a**) Detailed structure of the implemented MASH DDSM. (**b**) Level k which consists of N_k-bit EFMs. Modulus $M_k = 2^{N_k} = 2^6$ for level 1 to level 4 and $M_5 = M_6 = 2$ for level 5 and level 6

The MASH structure is implemented in levels, as shown in Fig. 7.3a [49]. The EFM in each level feeds its output to the level above. The structure of each level is shown in Fig. 7.3b, with the block diagram of the EFM1 and the configurable EFM shown in Fig. 7.4a and b. The top level outputs correspond to the outputs of each stage and they enter the noise cancellation network to generate the output of the MASH. The top four levels are 6-bit fixed-modulus EFMs and the bottom two levels are 3-bit variable-modulus EFMs. During the measurements, the bottom two levels

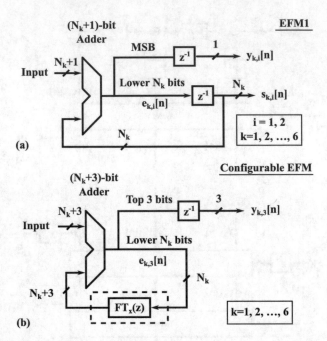

Fig. 7.4 The schematics of (**a**) the EFM1 and (**b**) the configurable EFM. $FT_x(z) = DT(z) + z^{-1}$ is the variable transfer function

are functioning as 1-bit EFMs with a modulus of 2. Thus, the divider controllers are 26 bits wide, i.e., the modulus is $M = 2^{26}$. In the implementation, the delayed output $y_{k,i}[n]$ is fed into the next level above, as shown in Fig. 7.4a and b. Delays are also added for the inputs of the first stage EFMs, as illustrated in the dashed box in Fig. 7.3a. Since the delayed quantization error $s_{k,i}[n]$ is injected into the next stage EFM1 in a level, as shown in Fig. 7.3b, delays are inserted for the outputs of the top level for correct cancellation. For the MASH 1-1-1 and its modified variants, dither $d_1[n]$ is injected from the input of the second stage EFM at the last level. The full-range dither $d_2[n]$ is applied to the input of each third stage EFM and the bit width of $d_{k,i}[n]$ corresponds to the bit width of the EFM.

In Mode 0, the MASH DDSM operates as a standard MASH 1-1-1, i.e., $DT(z) = 0$; this is the reference configuration for comparison purposes. Five first-order LSB dithered MASH 1-1-1-based wandering spur solutions are implemented and are denoted Mode 1 to Mode 5:

Mode 1: a $d_2[n]$ with a range of $[0, M - 1]$, $DT(z) = 0$,
Mode 2: modified third EFM1 with $DT(z) = -z^{-1} + 2z^{-2}$,
Mode 3: modified third EFM1 with $DT(z) = z^{-2}$,
Mode 4: modified third EFM1 with $DT(z) = -2z^{-1}(1 - z^{-1})$,
Mode 5: modified third EFM1 with $DT(z) = 2z^{-2}(1 - z^{-1})$.

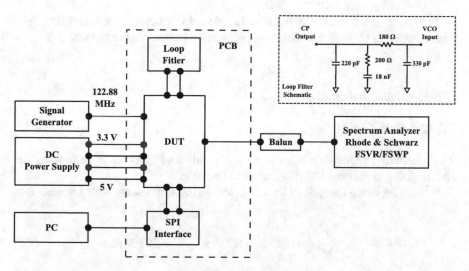

Fig. 7.5 Measurement setup for the fractional-N frequency synthesizer with MASH-based wandering spur solutions

7.1.3 Measurement Setup

The measurement setup for the fractional-N frequency synthesizer with the MASH-based wandering spur solutions is shown in Fig. 7.5. A passive RC loop filter that gives a bandwidth of 100 kHz is implemented externally using discrete components. The schematic of the loop filter is also shown in Fig. 7.5. A reference frequency of $f_{PFD} = 122.88$ MHz is used for all measurements. Bleed current is normally used to offset the operating point of the charge pump and the linearity of the region of operation changes with the operating point. The bleed current is chosen so that the PFD/CP nonlinearity is the *worst* case for the observation of the wandering spur patterns. During the measurement of the output phase noise profile, the bleed current is adjusted to provide the *best* PFD/CP linearity.

7.2 Performance of Wandering Spur Mitigation

In this section, measurement results are presented for each divider controller architecture. Spectrograms are shown for representative inputs that lead to prominent wandering spurs in the three cases, namely:

Case I, the input is $X = 1$,
Case II, the input is $X = M/4 + 1$,
Case III, the input is $X = M/2$ and the first stage initial condition is $s_1[0] = 1$.

In Case I and Case II the initial conditions do not affect the wandering spur pattern while in Case III the initial condition $s_1[0]$ influences the pattern.

7.2.1   Standard MASH 1-1-1

7.2.1.1   Case I

The standard MASH 1-1-1 leads to a typical wandering spur pattern in the spectrogram, as shown in Fig. 7.6, when the input is $X = 1$.

According to the analysis in Sect. 5.3, wandering spur events with a wander rate of

$$WR = f_{PFD}^2 \frac{X}{M} = (122.88 \times 10^6)^2 \times \frac{1}{2^{26}} = 225 \text{ MHz/s} \tag{7.1}$$

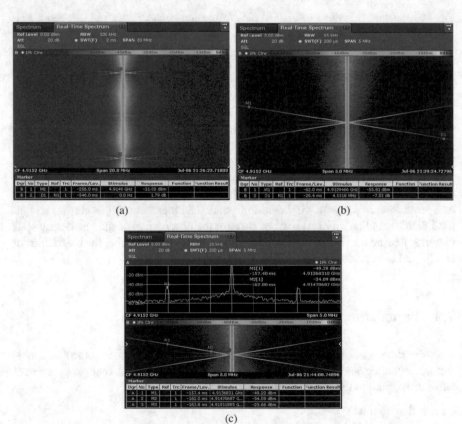

Fig. 7.6 (a) Measured output spectrogram of the synthesizer with the MASH 1-1-1, (b) zoom of a single wandering spur event, and (c) real-time spectrum showing the spur amplitude. Input $X = 1$ (Case I), $M = 2^{26}$, and $f_{PFD} = 122.88$ MHz

are expected to occur with a period of

$$T_{WS} = \frac{1}{f_{PFD}} \frac{M}{X} = \frac{1}{122.88 \times 10^6} \times \frac{2^{26}}{1} = 546.13 \text{ ms.} \tag{7.2}$$

In Fig. 7.6a, the delta marker indicates that the period of the events is $T_{WS} = 546$ ms.

The delta marker in Fig. 7.6b indicates that two points on the trace of the wandering spurs have a frequency difference Δf of 4.5318 MHz and a time difference Δt of 20.43 ms. This gives a measured wander rate of

$$WR = \frac{\Delta f}{\Delta t} = \frac{4.5318 \times 10^6}{20.43 \times 10^{-3}} = 222.15 \text{ MHz/s.} \tag{7.3}$$

These results match the predictions within the accuracy of the instrument.

7.2.1.2   Case II

The wandering spur pattern in the output spectrogram for a MASH 1-1-1-based synthesizer with input $X = M/4 + 1$ is shown in Fig. 7.7. The analysis in Sect. 5.4 predicts that the visible events should have a wander rate of

$$WR = f_{PFD}^2 \frac{|X'|}{M} = (122.88 \times 10^6)^2 \times \frac{1}{2^{26}} = 225 \text{ MHz/s} \tag{7.4}$$

and a period of

$$T_{WS} = \frac{1}{f_{PFD}} \frac{M}{D|X'|} = \frac{1}{122.88 \times 10^6} \times \frac{2^{26}}{4 \times 1} = 136.53 \text{ ms.} \tag{7.5}$$

In Fig. 7.7a, the delta marker indicates that the period of the events is 138 ms. The delta marker on the trace of spurs in Fig. 7.7b suggests a frequency difference Δf of 4.5443 MHz and a time difference Δt of 20.4 ms, giving a wander rate of 222.76 MHz/s. These results match the parameters of the expected pattern.

7.2.1.3   Case III

Figure 7.8 shows the wandering spur pattern in Case III when the input is $X = M/2$ and $s_1[0] = 1$. The spectrogram shows a complicated pattern in which multiple events at different wander rates can be observed, as analyzed in Sect. 5.5. The fundamental wander rate is predicted to be

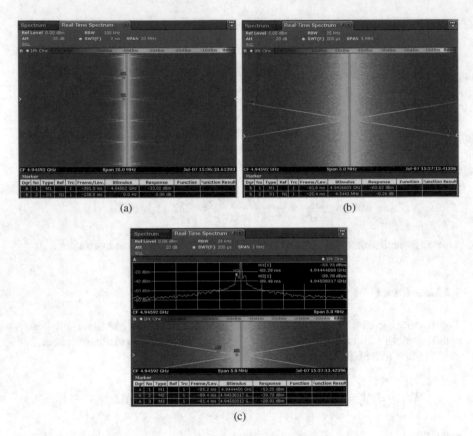

Fig. 7.7 (**a**) Measured output spectrogram of the synthesizer with a MASH 1-1-1, (**b**) zoom of a single wandering spur event, and (**c**) real-time spectrum showing the spur amplitude. Input $X = M/4 + 1 = 16777217$ (Case II), $M = 2^{26}$, and $f_{PFD} = 122.88$ MHz

$$WR = f_{PFD}^2 \left(\frac{s_1[0] + \mathbf{E}(d[n])}{M} \right)$$

$$= (122.88 \times 10^6)^2 \times \frac{1 + 0.5}{2^{26}}$$

$$= 337.5 \text{ MHz/s}. \tag{7.6}$$

The delta marker indicates a frequency difference Δf of 4.6005 MHz and a time difference Δt of 13.6 ms, giving a wander rate of 338.27 MHz/s. This matches the analytically predicted wander rate.

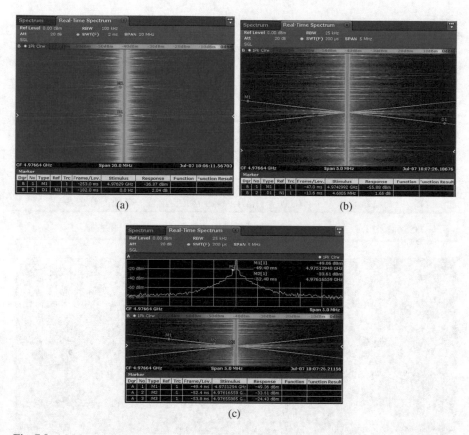

(a)

(b)

(c)

Fig. 7.8 (**a**) Measured output spectrogram of the synthesizer with a MASH 1-1-1, (**b**) zoom of a single wandering spur event, and (**c**) real-time spectrum showing the spur amplitude. Input $X = M/2 = 33554432$ (Case III), $s_1[0] = 1$, $M = 2^{26}$, and $f_{PFD} = 122.88$ MHz

7.2.2 Summary of Wandering Spur Patterns Observed in the MASH 1-1-1-Based Synthesizer

A summary of the predicted and measured fundamental wander rate and period in each case is shown in Table 7.1.

7.2.3 Mode 1: Uniform Dither Solution

7.2.3.1 Case I

Figure 7.9 shows the measured output spectrograms of the synthesizer in Mode 1 with the input $X = 1$. The markers in Fig. 7.9b show that the spur amplitudes are -33.33 dBm, -43.98 dBm, and -56.62 dBm at offsets of about 70 kHz,

Table 7.1 Predicted and measured wandering spur patterns of MASH 1-1-1

Case I and case II

Wandering spur case	Input (X)	Δf (MHz)	Δt (ms)	Predicted wander rate (MHz/s)	Measured wander rate (MHz/s)	Predicted period (ms)	Measured period (ms)
Case I	1	4.5318	20.4	225	222.15	546.13	546
Case II	$M/4+1 =$ 16777217	4.5443	20.4	225	222.76	136.53	138

Case III

Input (X)	Initial condition $(s_1[0])$	Δf (MHz)	Δt (ms)	Predicted wander rate (MHz/s)	Measured wander rate (MHz/s)
$\dfrac{M}{2} = 33554432$	1	4.6005	13.6	337.50	338.27

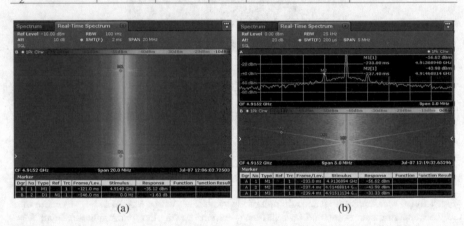

(a)   (b)

Fig. 7.9 (**a**) Measured output spectrogram and (**b**) real-time spectrum of synthesizer with the modified MASH in Mode 1. Input $X = 1$ (Case I), $M = 2^{26}$, and $f_{PFD} = 122.88$ MHz

500 kHz, and 1.5 MHz from the carrier, respectively. The markers at similar offsets in Fig. 7.6c report spur amplitudes of -23.66 dBm, -34.09 dBm, and -49.20 dBm. Thus, the divider controller in Mode 1 does not eliminate the wandering spurs; nevertheless, it reduces the wandering spur amplitudes by about 10 dB.

7.2.3.2 Case II

Figure 7.10 shows the measured output spectrograms of the synthesizer with its divider controller in Mode 1 when the input is $X = M/4 + 1 = 16777217$. Comparing the spur amplitudes given by the markers in Fig. 7.10b (-37.49 dBm, -48.33 dBm, and -61.49 dBm) and those in Fig. 7.7c (-28.81 dBm, -39.78 dBm,

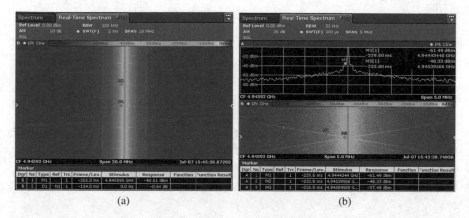

(a)   (b)

Fig. 7.10 (**a**) Measured output spectrogram and (**b**) the real-time spectrum of the synthesizer with the modified MASH in Mode 1. Input $X = M/4 + 1 = 16777217$ (Case II), $M = 2^{26}$, and $f_{PFD} = 122.88$ MHz

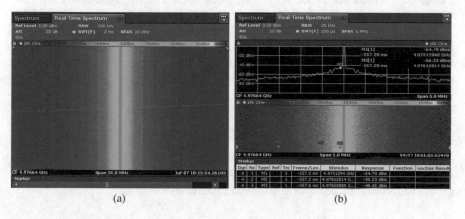

(a)   (b)

Fig. 7.11 (**a**) Measured output spectrogram and (**b**) the real-time spectrum of the synthesizer with the modified MASH in Mode 1. Input $X = M/2 = 33554432$ (Case III), $s_1[0] = 1$, $M = 2^{26}$, and $f_{PFD} = 122.88$ MHz

and -53.25 dBm), Mode 1 can reduce the wandering spur amplitudes by about 8.5 dB in Case II.

7.2.3.3   Case III

Figure 7.11 shows the measured output spectrograms of the synthesizer in Mode 1. The input is $X = M/2 = 33554432$ and the initial condition of the first stage is $s_1[0] = 1$. No wandering spur pattern is observed in this case.

7.2.4   Mode 2: Modified Third EFM1 with $DT(z) = -z^{-1} + 2z^{-2}$

7.2.4.1   Case I

The measured output spectrograms when the synthesizer is in Mode 2 are shown in Fig. 7.12. The input is $X = 1$. Comparing the amplitude information of the markers with similar offset frequencies on the wandering spur trace in Fig. 7.12b (-40.34 dBm, -52.75 dBm, and -64.04 dBm) and Fig. 7.6c (-23.66 dBm, -34.09 dBm, and -49.20 dBm), Mode 2 mitigates the wandering spurs by more than 15 dB in Case I.

7.2.4.2   Case II

Figure 7.13 shows the measured output spectrograms of the synthesizer with the divider controller in Mode 2 and its input is $X = M/4 + 1 = 16777217$. A comparison between markers at similar offsets from the carrier in Fig. 7.13b (-46.16 dBm, -56.08 dBm, and -66.74 dBm) and Fig. 7.7c (-28.81 dBm, -39.78 dBm, and -53.25 dBm) indicates that Mode 2 can reduce the wandering spur amplitudes by more than 13 dB.

7.2.4.3   Case III

Figure 7.14 shows the spectrograms of the output of the synthesizer with Mode 2 enabled. The input is $X = M/2 = 33554432$ and the first stage initial condition is $s_1[0] = 1$. No wandering spur is observed in these spectrograms.

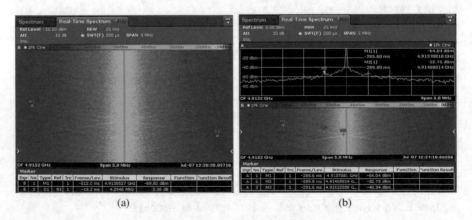

(a)   (b)

Fig. 7.12 (a) Measured output spectrogram and (b) the real-time spectrum of synthesizer with the modified MASH in Mode 2. Input $X = 1$ (Case I), $M = 2^{26}$, and $f_{PFD} = 122.88$ MHz

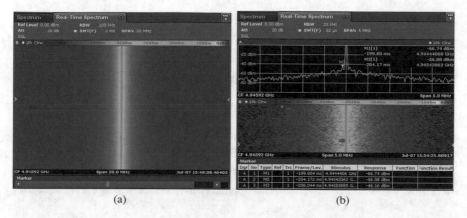

(a)   (b)

Fig. 7.13 (a) Measured output spectrogram and (b) the real-time spectrum of synthesizer with the modified MASH in Mode 2. Input $X = M/4 + 1 = 16777217$ (Case II), $M = 2^{26}$, and $f_{PFD} = 122.88$ MHz

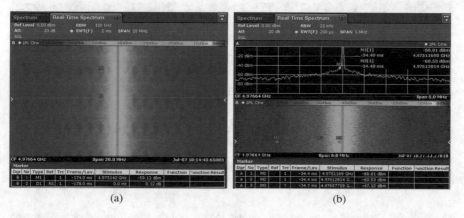

(a)   (b)

Fig. 7.14 (a) Measured output spectrogram and (b) the real-time spectrum of synthesizer with the modified MASH in Mode 2. Input $X = M/2 = 33554432$ (Case III), $s_1[0] = 1$, $M = 2^{26}$, and $f_{PFD} = 122.88$ MHz

7.2.5   Mode 3: Modified Third EFM1 with $DT(z) = z^{-2}$

7.2.5.1   Case I

Figure 7.15 shows the measured output spectrograms of the synthesizer in Mode 3. The divider controller input is $X - 1$. No wandering spur pattern is visible, indicating that this solution eliminates the wandering spur in Case I.

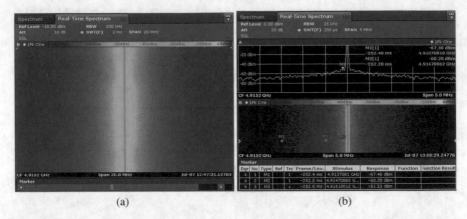

(a)   (b)

Fig. 7.15 (a) Measured output spectrogram and (b) the real-time spectrum of synthesizer with the modified MASH in Mode 3. Input $X = 1$ (Case I), $M = 2^{26}$, and $f_{PFD} = 122.88$ MHz

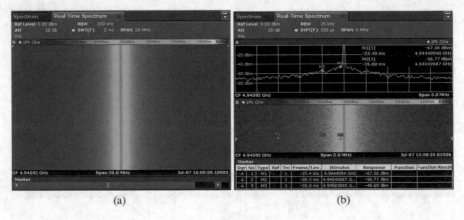

(a)   (b)

Fig. 7.16 (a) Measured output spectrogram and (b) the real-time spectrum of synthesizer with the modified MASH in Mode 3. Input $X = M/4 + 1 = 16777217$ (Case II), $M = 2^{26}$, and $f_{PFD} = 122.88$ MHz

7.2.5.2   Case II

The measured output spectrograms when the synthesizer in Mode 3 with a divider controller input $X = M/4 + 1 = 16777217$ are shown in Fig. 7.16. The wandering spur pattern is not seen in the output spectrogram in this case.

7.2.5.3   Case III

Figure 7.17 shows the spectrograms of the output when the synthesizer is in Mode 3. The input of the MASH is $X = M/2 = 33554432$ and the first stage initial condition is $s_1[0] = 1$. No wandering spur is observed in the spectrograms.

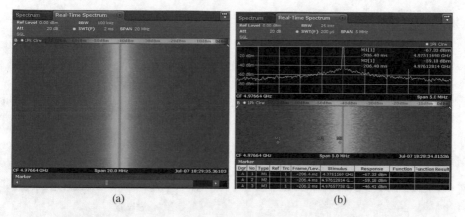

(a)   (b)

Fig. 7.17 (**a**) Measured output spectrogram and (**b**) the real-time spectrum of synthesizer with the modified MASH in Mode 3. Input $X = M/2 = 33554432$ (Case III), $s_1[0] = 1$, $M = 2^{26}$, and $f_{PFD} = 122.88$ MHz

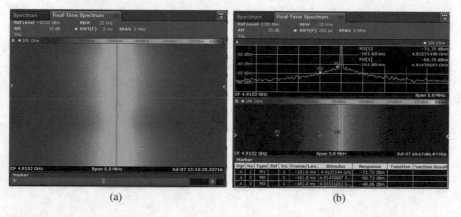

(a)   (b)

Fig. 7.18 (**a**) Measured output spectrogram and (**b**) the real-time spectrum of synthesizer with the modified MASH in Mode 4. Input $X = 1$ (Case I), $M = 2^{26}$, and $f_{PFD} = 122.88$ MHz

It can be concluded that Mode 3 successfully eliminates wandering spurs in all three cases.

7.2.6 Mode 4: Modified Third EFM1 with $DT(z) = -2z^{-1}(1 - z^{-1})$

7.2.6.1 Case I

The measured output spectrograms of the synthesizer with Mode 4 enabled are shown Fig. 7.18. The MASH input is $X = 1$. No wandering spur pattern is visible, suggesting that Mode 4 can eliminate wandering spurs in Case I.

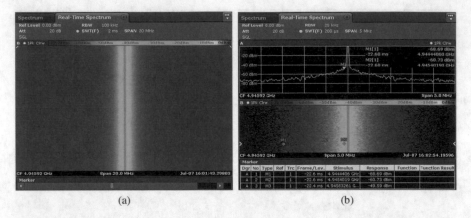

Fig. 7.19 (a) Measured output spectrogram and (b) the real-time spectrum of synthesizer with the modified MASH in Mode 4. Input $X = M/4 + 1 = 16777217$ (Case II), $M = 2^{26}$, and $f_{PFD} = 122.88$ MHz

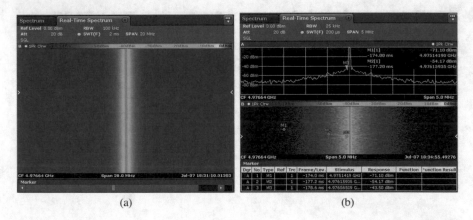

Fig. 7.20 (a) Measured output spectrogram and (b) the real-time spectrum of synthesizer with the modified MASH in Mode 4. Input $X = M/2 = 33554432$ (Case III), $s_1[0] = 1$, $M = 2^{26}$, and $f_{PFD} = 122.88$ MHz

7.2.6.2 Case II

Figure 7.19 shows the measured output spectrograms of Case II when the divider controller is in Mode 4 and input is $X = M/4 + 1 = 16777217$. The wandering spur pattern is not observed in the output spectrogram.

7.2.6.3 Case III

Figure 7.20 shows the spectrogram of the output in Case III when the divider controller is in Mode 4 with an input $X = M/2 = 33554432$ and a first stage initial condition $s_1[0] = 1$. A barely visible wandering spur pattern is observed in the spectrogram. Comparing the spur amplitudes at corresponding markers in Fig. 7.20b (-43.50 dBm, -54.17 dBm, and -71.10 dBm) and Fig. 7.8c (-24.43 dBm, -33.61 dBm, and -49.06 dBm), Mode 4 reduces the wandering spur by at least 20 dB.

7.2.7 Mode 5: Modified Third EFM1 with $DT(z) = 2z^{-2}(1 - z^{-1})$

7.2.7.1 Case I

Figure 7.21 shows the measured output spectrograms of the synthesizer when the divider controller is in Mode 5 with a Case I input $X = 1$. No wandering spur pattern is visible in the spectrograms.

7.2.7.2 Case II

The measured output spectrograms when the synthesizer in Mode 5 with a Case II input $X = M/4 + 1 = 16777217$ are shown in Fig. 7.22. The output spectrograms do not show any wandering spur pattern.

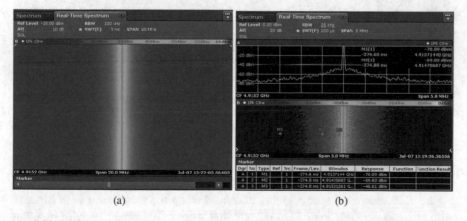

(a)   (b)

Fig. 7.21 (a) Measured output spectrogram and (b) the real-time spectrum of synthesizer with the modified MASH in Mode 5. Input $X = 1$ (Case I), $M = 2^{26}$, and $f_{PFD} = 122.88$ MHz

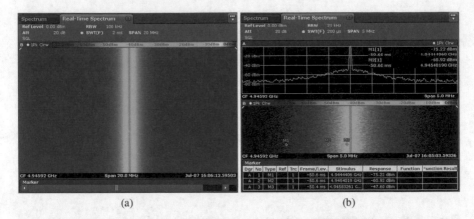

Fig. 7.22 (**a**) Measured output spectrogram and (**b**) the real-time spectrum of synthesizer with the modified MASH in Mode 5. Input $X = M/4 + 1 = 16777217$ (Case II), $M = 2^{26}$, and $f_{PFD} = 122.88$ MHz

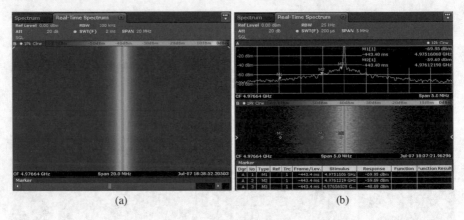

Fig. 7.23 (**a**) Measured output spectrogram and (**b**) the real-time spectrum of synthesizer with the modified MASH in Mode 5. Input $X = M/2 = 33554432$ (Case III), $s_1[0] = 1$, $M = 2^{26}$, and $f_{PFD} = 122.88$ MHz

7.2.7.3   Case III

Figure 7.23 shows the spectrograms of output when the synthesizer is in Mode 5 with an input $X = M/2 = 33554432$ and first stage initial condition $s_1[0] = 1$. No wandering spur is observed in the spectrograms.

In conclusion, Mode 5 successfully eliminates wandering spurs in all three cases.

Table 7.2 Summary of the performance of wandering spur solutions

Divider controller mode	Performance in representative cases		
	Case I	Case II	Case III
Mode 1	About 10 dB reduction	About 8.5 dB reduction	No pattern observed
Mode 2	More than 15 dB reduction	More than 13 dB reduction	No pattern observed
Mode 3	No pattern observed	No pattern observed	No pattern observed
Mode 4	No pattern observed	No pattern observed	At least 20 dB reduction
Mode 5	No pattern observed	No pattern observed	No pattern observed

7.2.8   Summary of the Performance of Wandering Spur Mitigation

Table 7.2 summarizes the effectiveness of the implemented wandering spur solutions. Mode 1 and Mode 2 cannot eliminate the wandering spurs but the divider controller structures can reduce the wandering spurs amplitudes in all cases. Mode 4 shows a very weak pattern of wandering spurs in Case III. The divider controller in Mode 3 and Mode 5 can eliminate the wandering spurs in all three representative cases.

7.3   Output Phase Noise Performance

In this section, measurements of the output phase noise performance of the synthesizer with different divider controller modes enabled are presented. Estimates for the output phase noise are given. In the estimates, the reference oscillator noise is modeled as a white noise of -152 dBc/Hz. The VCO noise is modeled by the sum of a $1/f^2$ noise of -136 dBc/Hz at 1 MHz offset and a white noise of -164 dBc/Hz. To model the flicker noise at low offset frequencies, an input-referred $1/f$ noise of -135 dBc/Hz at 1 kHz offset is considered.

7.3.1   Standard MASH 1-1-1 and the Uniform Dither Solution

Figure 7.24 shows the phase noise profile of a standard MASH 1-1-1 and the uniform dither wandering spur solution. The jitters of the synthesizer were measured to be 49.4 fs and 91.9 fs, respectively, when these two modes were enabled.

The Mode 1 uniform dither solution has a noise profile of a second-order divider controller. The ripples in the phase noise of Mode 1 can be related to the fact that the

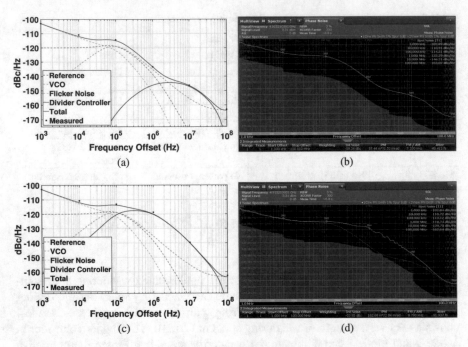

Fig. 7.24 Estimated and measured output phase noise spectrum of the synthesizer (**a**), (**b**) with a MASH 1-1-1 and (**c**), (**d**) in Mode 1. Input $X = 5461$, $M = 2^{26}$, and $f_{PFD} = 122.88$ MHz

phase noise contribution in Mode 1 is dominated by the additional dither $d_2[n]$. The performance of the pseudorandom number generator affects the phase noise profile.

7.3.2 Modified MASH Solutions

The output phase noise profiles of the modified MASH solutions, including second-order solutions (Mode 2 and Mode 3) and third-order solutions (Mode 4 and Mode 5), are shown in Figs. 7.25 and 7.26. Notice that the solutions exhibit a typical noise profile for a second-order divider controller for Mode 2 and Mode 3, leading to the high noise at mid-range frequency offsets. The additional high-frequency contributions of the Mode 4 and Mode 5 divider controllers also increase the total phase noise when compared with the standard MASH 1-1-1-based synthesizer. The jitters for the synthesizer in Mode 2, Mode 3, Mode 4 and Mode 5 are 96.5 fs, and 93.8 fs, 58.0 fs, and 56.8 fs, respectively.

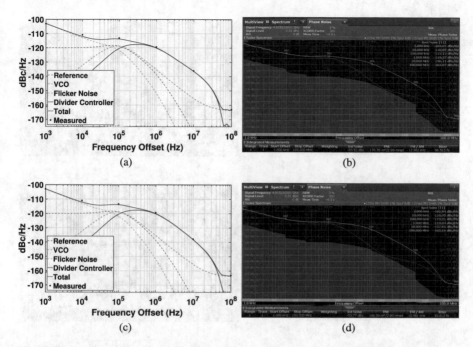

Fig. 7.25 Estimated and measured output phase noise spectrum of the synthesizer with (**a**), (**b**) modified MASH in Mode 2, (**c**), (**d**) Mode 3. Input $X = 5461$, $M = 2^{26}$, and $f_{PFD} = 122.88$ MHz

7.3.3 Summary of Jitter Performance

The jitters of the synthesizer with different divider controller modes enabled are summarized in Table 7.3. The second-order designs (Mode 2 and Mode 3) increase the jitter by about 40 fs while the third-order designs (Mode 4 and Mode 5) increase it by about 8 fs, when compared with the 49 fs jitter of the reference synthesizer with a MASH 1-1-1 (Mode 0).

7.3.4 IBS Performance

As discussed in Chap. 6, the wandering spur solutions may also reduce the amplitudes of fixed spurs. Integer boundary spur measurement results for the synthesizer with the standard MASH 1-1-1 and the Mode 5 wandering spur solution are shown in Fig. 7.27. The results indicate that Mode 5 can achieve an improvement of about 2 dB in IBS amplitude. This is consistent with the observation in Sect. 6.4.

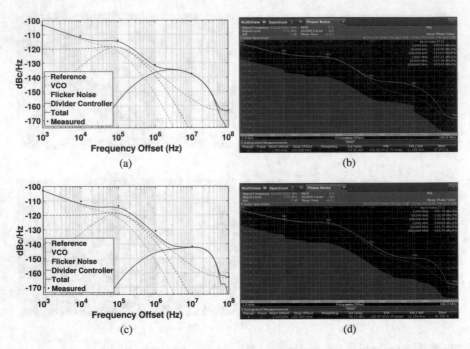

Fig. 7.26 Estimated and measured output phase noise spectrum of the synthesizer with (**a**), (**b**) modified MASH in Mode 4, (**c**), (**d**) Mode 5. Input $X = 5461$, $M = 2^{26}$, and $f_{PFD} = 122.88$ MHz

Table 7.3 Summary of measured jitter performance (1 kHz to 100 MHz)

Divider controller mode	Mode 0	Mode 1	Mode 2	Mode 3	Mode 4	Mode 5
Jitter (fs)	49.4	91.9	96.5	93.8	58.0	56.8

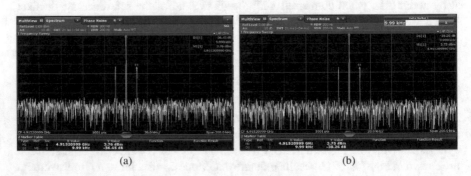

Fig. 7.27 Measured output spectrum of synthesizers with (**a**) the standard MASH 1-1-1 and (**b**) modified MASH in Mode 5. Input $X = 5461$, $M = 2^{26}$, and $f_{PFD} = 122.88$ MHz. Bleed current is set to 0, corresponding to a highly nonlinear region of the charge pump transfer characteristic

7.4 Summary

The implemented fractional-N frequency synthesizer with the 26-bit divider controller configured as a MASH 1-1-1 manifests wandering spurs, as predicted.

The effectiveness of various wandering spur mitigation techniques has been evaluated by observing spectrograms and real-time spectra of the synthesizer output phase noise. As summarized in Table 7.2, the uniform dither solution is the least effective; it can mitigate but not eliminate wandering spurs in all cases. Mode 2 with $DT(z) = -z^{-1} + 2z^{-2}$ is a more effective second-order solution in comparison. Mode 3 with $DT(z) = z^{-2}$ is also a second-order solution and it eliminates wandering spurs in all three cases. The third-order solution Mode 4 with $DT(z) = 2z^{-2}(1 - z^{-1})$ leads to no wandering spur pattern in Case I or Case II. However, the synthesizer in this mode shows a very weak wandering spur pattern in Case III. The other third-order solution Mode 5 with $DT(z) = 2z^{-2}(1 - z^{-1})$ is an effective design that eliminates wandering spurs in all three cases.

A summary of the jitter performance of the synthesizer in the five modes is presented in Table 7.3. The synthesizer with second-order solutions enabled has significantly more noise at mid-range offset frequencies and consequently the jitter performance is about 40 fs worse than the standard MASH 1-1-1. The third-order solutions have higher contributions to the phase noise at higher offset frequencies but they can achieve jitter performance that is only about 8 fs worse than the standard MASH design. For this reason, among the two designs that are most effective, Mode 5 is a better option than Mode 3 because of its lower jitter penalty.

A comparison of the IBS performance between the standard MASH 1-1-1 and the MASH in Mode 5 indicates that the Mode 5 MASH reduces the fundamental IBS by about 2 dB when the PFD/CP linearity is at its worst.

Appendix A
Relation Between Fractional Boundary Spurs and Fractional Spurs

In Chap. 3, three types of MASH 1-1-1 DDSM-induced spurs are discussed. Fractional boundary spurs appear when the input can be expressed as

$$X = \frac{X_{Num}}{X_{Den}} M = \frac{aM}{b} + \frac{cM}{D}, \tag{A.1}$$

where the N-bit MASH 1-1-1 has a modulus $M = 2^N$.[1] Here X_{Num} and X_{Den}, a and b, and c and D are pairs of coprime integers, while $D \gg b$.

Assume the product of all odd factors of b is b_o, then

$$b = 2^k b_o. \tag{A.2}$$

The input, which is an integer, can be expressed as

$$X = \frac{aM}{b} + \frac{cM}{D} \tag{A.3}$$

$$= \frac{a2^{N-k} + cMb_o/D}{b_o}. \tag{A.4}$$

If $LCM(b, D) \neq D$, then cMb_o/D is a fractional number and thus the input is not an integer. To make sure that the numerator is an integer, D should have the form

$$D = 2^{k+q} b_o, \tag{A.5}$$

---

[1] It is assumed that modulus M is an integer power of two. Similar conclusions apply to other integer values of M.

© The Author(s), under exclusive license to Springer Nature Switzerland AG 2022
D. Mai, M. P. Kennedy, *Wandering Spurs in MASH-based Fractional-N Frequency Synthesizers*, Analog Circuits and Signal Processing,
https://doi.org/10.1007/978-3-030-91285-7

where q is an integer. This means

$$LCM(b, D) = D. \tag{A.6}$$

In the conventional definition [31], fractional spurs are the spurs at offset frequencies of

$$f_k = \frac{k}{X_{Den}} f_{PFD}, \quad k = 1, 2, \ldots. \tag{A.7}$$

The simple fraction expression of the input can be written as

$$X = \frac{aM}{b} + \frac{cM}{D} = \frac{\frac{a}{b}D + c}{D} M \tag{A.8}$$

$$= \frac{a2^q + c}{D} M = \frac{X_{Num}}{X_{Den}} M \tag{A.9}$$

The fundamental of the fractional boundary spurs appears at an offset frequency

$$f_{FBS} = \frac{bc}{D} f_{PFD} \tag{A.10}$$

$$= \frac{c}{D/b} f_{PFD} = \frac{c}{2^q} f_{PFD}. \tag{A.11}$$

The possible cases can be classified based on the parities of b and D. When b is even and D is even, both a and c are odd integers. Hence the sum $a2^q + c$ is an odd integer. Since b_o contains all odd factors of D, $D/b = 2^q \leq X_{Den}$.

When b is odd, D can be even. In this case, $a2^q + c$ is odd. Again, according to (A.5), $D/b = 2^q \leq X_{Den}$.

When b and D are both odd, it can be deduced that $D = b = b_o$. This contradicts the assumption that $D \gg b$, meaning that this case does not arise.

Therefore, the fractional boundary spurs are fractional spurs according to the conventional definition. Also, sub-harmonic spurs can be related to sub-fractional spurs. Sub-harmonic spurs appear when the input in the form of (A.1) with $b = 2^x$ and $D = 2^y$, i.e.,

$$X = \frac{aM}{2^x} + \frac{cM}{2^y} \tag{A.12}$$

and the offsets of the spurs are

$$f_{sub,k} = \frac{(2k + 1)}{2D} f_{PFD}, \quad k = 0, 1, 2, \ldots \tag{A.13}$$

The definition of the sub-fractional spurs refers to the spurs typically at

$$f_{k,l} = \frac{k}{lX_{Den}} f_{PFD}, \quad k = 1, 2, \ldots, l = 2, 3, \ldots. \tag{A.14}$$

Comparing (A.13) and (A.14), it can be further concluded that sub-harmonic spurs are sub-fractional spurs.

It should be noticed that the offset frequencies of the sub-harmonic spurs depend on the initial condition. More generally, the spur offset frequencies can be expressed by

$$\left(\frac{(2k+1)}{2D} + \alpha \right) f_{PFD}, \quad k = 0, 1, 2, \ldots, |\alpha| \leq \frac{1}{2D}. \tag{A.15}$$

The definitions of spurs mentioned in this book are summarized in Table A.1. A Venn diagram illustrating the types of MASH DDSM-induced fixed spurs is shown in Fig. A.1.

Table A.1 Summary of spur definitions

Input			
$X = \dfrac{X_{Num}}{X_{Den}}$ $M = \dfrac{aM}{b} + \dfrac{cM}{D}$, $M = 2^N$			
Spur type	Offset frequencies		
Integer boundary spurs	$k\dfrac{X_{Num}}{X_{Den}} f_{PFD} = k\dfrac{c}{D} f_{PFD}$, $k = 1, 2, \ldots; a = 0$ or 1, $b = 1$		
Spur type, from this book	Offset frequencies	Spur type, conventional [31]	Offset frequencies
Fractional boundary spurs	$k\dfrac{bc}{D} f_{PFD}$, $k = 1, 2, \ldots;$ $a \neq 0$, $b \geq 2$	Fractional spurs	$\dfrac{k}{X_{Den}} f_{PFD}$, $k = 1, 2, \ldots$
Sub-harmonic spurs	$\left(\dfrac{(2k+1)}{2D} + \alpha \right) f_{PFD}$, $k = 0, 1, 2, \quad |\alpha| \leq \dfrac{1}{2D}$; $b = 2^x$, $D = 2^y$	Sub-fractional spurs	$\dfrac{k}{lX_{Den}} f_{PFD}$, $k = 1, 2,$ $l = 2, 3, \ldots$

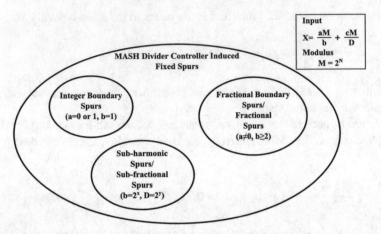

Fig. A.1 Venn diagram illustrating the relation between the MASH DDSM-induced fixed spurs

Appendix B
Proofs of Theorems

B.1 Proof of Theorem 5.2

Proof The difference between terms of the sequence $T_1'[n]$ is

$$\frac{DT_1[n-1+N]}{M} - \frac{DT_1[n-1]}{M} = \frac{(2n+1+N)Nk}{2}. \tag{B.1}$$

If $((l + D/2) \bmod D)$ exists in $T_1'[n]$, then

$$\frac{(2n+1+N)Nk}{2} = \frac{D}{2} + pD. \tag{B.2}$$

This leads to

$$(2n+1+N)Nk = (2p+1)D. \tag{B.3}$$

When $N = D$, $(2n + 1 + D)k$ is odd because D is even. Therefore

$$p = \frac{(2n+1+D)k-1}{2}. \tag{B.4}$$

This means that $(l + D/2) \bmod D$ exists in $T_1'[n]$ and it is D samples away from the term that is equal to l. Following Theorem 5.1, $T_1'[n]$ can be found to be symmetrical within a $2D$ period. It can be deduced that, in the even D case, D pairs of terms with a difference of $D/2$ under the modulo operation exist in a $\Delta = 2D$ period in $T_1'[n]$. This indicates that D pairs of terms with a difference of approximately $M/2$ under the modulo operation can exist in a $\Delta = 2D$ period in $e_2[n]$. □

B.2   Proof of Theorem 5.3

Proof The difference between adjacent terms in $T_1'[n]$ is

$$\frac{DT_1[n]}{M} - \frac{DT_1[n-1]}{M} = (n+1)k. \tag{B.5}$$

Consider at $n = m_0$, $(m_0 + 1)k \equiv 0 \pmod{D}$ and $T_1'[m_0] = 0$. Because k is odd in cases of even D, the terms in the sequence of differences have alternating parities. For the next D terms in $DT_1[n-1]/M$, the sum of any number of consecutive differences is

$$\frac{(m_0 + 1 + a)k + (m_0 + 1 + b)k}{2}(b - a + 1),$$

$$1 \le a \le b \le D - 1 \tag{B.6}$$

and it is impossible that this sum is an integer multiple of D when $D = 2^p$. Consider two possible cases:

1. $(m_0 + 1 + a)k$ and $(m_0 + 1 + b)k$ have different parities and their sum is odd. Since the parities of the terms are alternating, $(b - a + 1)$ is even. Because D does not contain any odd factor besides unity and $0 < (b - a + 1) < D$, in this case the sum cannot be an integer multiple of D.
2. $(m_0 + 1 + a)k$ and $(m_0 + 1 + b)k$ have the same parity and their sum is even. In this case $(b - a + 1)$ is odd. The sum (B.6) modulo D is then

$$\frac{(a + b)k}{2}(b - a + 1) \bmod D \ne 0 \tag{B.7}$$

because $(a + b)/2 < D$.

In conclusion, $T_1'[n]$ will have D distinct values i.e. from 0 to $D - 1$ from $n = m_0 + 1$ on. Therefore, in the case where $D = 2^p$, the $e_2[n]$ sequence can appear to be equally-spaced by M/D.   □

B.3   Proof of Theorem 5.4

Proof Theorem 5.1 shows that $T_1'[n]$ is symmetrical within a period of $2D$. For the cases of an odd D, the period of $T_1[n]$ is D. Therefore, $T_1'[n]$ can be found mirrored within a period D, which indicates that $(D + 1)/2$ different values can exist in the sequence at most. Assume $D = D_1 D_2$ in this case, where D_1 and D_2 are two odd integers with D_2 being a prime factor less than D. The solutions to

$$\frac{n(n+1)k}{2} \equiv qD_2 + l \pmod{D}, \quad q = 0, \ 1 \cdots, \ D_1 - 1 \tag{B.8}$$

exist for all q values for a given integer l when $T_1'[n]$ is equally-spaced.

First consider the simplest case where D is a prime number. Therefore $D_2 = 1$ and at least D different values should be present if $T_1'[n]$ is equally-spaced. However, only $(D+1)/2$ different values exist in the sequence at most. The conclusion holds with $D_1 = 1$, meaning $D_2 > 1$.

A more complicated case is where D_1 and D_2 are both prime numbers greater than unity. Since

$$\frac{(n+\Delta)(n+1+\Delta)k}{2} - \frac{n(n+1)k}{2} = \frac{\Delta(2n+1+\Delta)k}{2}, \tag{B.9}$$

Equation (B.8) has an integer solution when $\Delta = D_2$ or integer multiples of D_2 for a given integer l. This means that terms that are D_2 samples apart in $T_1'[n]$ belong to the same set of solutions to (B.8) if $T_1'[n]$ is equally-spaced, i.e., $T_1'[n_0]$ and $T_1'[n_0 + kD_2]$ belong to the same set with an integer l.

Assume $T_1'[m_0] = 0$ and $(m_0 k) \bmod D = 0$. It can be shown that the first $(D_2 + 1)/2$ terms starting from $T_1'[m_0]$ belong to different sets of equally-spaced values. According to (B.5), the difference between any of two terms in $DT_1[n-1]/M$ from m_0 on is

$$\frac{((m_0+a)+(m_0+b))k}{2}(a-b+1), \quad 1 \le a \le b \le (D_2-1)/2. \tag{B.10}$$

We need to know if any of the differences is a multiple of D_2:

$$\frac{((m_0+a)+(m_0+b))k}{2}(a-b+1) \bmod D_2 \tag{B.11}$$

$$= \frac{(a+b)k}{2}(a-b+1) \bmod D_2. \tag{B.12}$$

Because $a + b < D_2$ and $a - b + 1 < D_2$, none of the differences is an integer multiple of D_2. Each of these terms belongs to a set of solutions of (B.8) with a distinct l.

Due to the aforementioned symmetry of $T_1'[n]$ within a period of D and the fact that terms D_2 samples apart belong to the same set of solutions to (B.8), $(D_2 + 1)/2$ different sets of equally-spaced values exist. This is true since the first $(D_2 + 1)/2$ terms starting from $T_1'[m_0]$ belong to different sets of equally-spaced values. Therefore, at least $D_1(D_2 + 1)/2$ different values exist in $T_1'[n]$. However, due to symmetry, only $(D+1)/2 = (D_1 D_2 + 1)/2$ different values can exist in $T_1'[n]$ at most. Thus, $T_1'[n]$ cannot be found equally-spaced with a prime D_2.

The set of terms starting with $T_1'[m_0 + (D_2 - 1)/2]$ corresponds to a distinct integer l since they begin with the $((D_2 + 1)/2)^{th}$ term starting at $T_1'[m_0]$. We denote the subsequence of $T_1'[n]$ starting from the $((D_2 + 1)/2)^{th}$ term from m_0 and equally spaced by D_2 samples as $T_{1,1}'[n]$. According to (B.9), the differences between the first term in $T_{1,1}'[n]$ and the following terms are

$$\frac{aD_2(2m_0 + (a + 1)D_2)k}{2} \equiv \frac{a(a + 1)}{2}D_2^2k \pmod{D}, \tag{B.13}$$

where a is an integer and this subset of $T_1'[n]$ follows a quadratic pattern too.

Now consider that D_1 is a composite number. Let $D_1 = D_3D_4$ where D_3 is a prime factor smaller than D_1. We focus on the distinct subset of the sequence $T_{1,1}'[n]$ with terms D_3 samples apart, which we denote as $T_{1,2}'[n]$, starting from the $((D_3 + 1)/2)^{th}$ term beginning at $T_{1,1}'[m_{0,1}]$, where $T_{1,1}'[m_{0,1}] - T_{1,1}'[m_{0,1} - 1] = 0$ and $(m_{0,1}kD_2^2) \bmod D = 0$. The differences between the first term and the following ones in $T_{1,2}'[n]$ are

$$\frac{b(b + 1)}{2}D_3^2D_2^2k \bmod D = \left(\frac{b(b + 1)}{2}kD_3D_2 \bmod D_4\right)D_2D_3. \tag{B.14}$$

If D_4 is a prime number or only contains two prime factors, based on the analysis before, $T_{1,2}'[n]$ is not equally-spaced and all sets including it are not equally-spaced as the result. If this is not the case, one could factorize a prime factor D_5 out of D_4 and continue the analysis of a subset of $T_{1,2}'[n]$. This can be repeated until D_n is prime or contains only two prime factors. Then, based on the prior analysis, all subsets involved are not equally-spaced.

This means that, for odd D cases where $D > 1$, $T_1'[n]$ does not have equally-spaced values and $e_2[n]$ is not equally-spaced around n_0 given by (5.41).   □

B.4   Proof About $e_2[n]$ Structure in Case II When D Is Even

In Sect. 5.4.3.1, it is shown that near the value of n_0 that gives $e_1[n_0] \approx pM/(2D)$, where p is odd, there is an $n_0' = n_0 + \delta$ such that

$$e_1[n_0'] = e_1[n_0 + \delta] \approx \frac{kM}{2D}. \tag{B.15}$$

The expression of $e_2[n]$ can be written as

$$e_2[n] \approx \left(e_2[n_0'] + \left((n - n_0 + 1)^2 - 1\right)\frac{kM}{2D}\right) \bmod M. \tag{B.16}$$

The pattern of the $e_2[n]$ waveform around n_0 should be determined by

$$\left(m^2 \frac{kM}{2D}\right) \bmod M. \tag{B.17}$$

The $e_2[n]$ expression in (B.16) is a shifted version of (B.17) with $m = n - n_0 + 1$.

The constant vertical shifting does not change the relative differences in amplitude under the modulo operation. Assume $x[n] = (x[n]) \bmod M$ is the original waveform. Based on the property of the modulo operation, for the vertically shifted version $(x[n] + C) \bmod M$, where C is a constant, we have

$$\left((x[n_2] + C) \bmod M - (x[n_1] + C) \bmod M\right) \bmod M = (x[n_2] - x[n_1]) \bmod M. \tag{B.18}$$

Therefore, if (B.17) does not exhibit an equally-spaced pattern, then it suggests that $e_2[n]$ around n_0 does not exhibit an equally-spaced pattern either. Next, the proof that (B.17) does not exhibit an equally-spaced pattern is given.

Proof The pattern of (B.17) is determined by the sequence

$$(n^2 k) \bmod 2D. \tag{B.19}$$

If the sequence is equally-spaced, it should be at least equally-spaced by D. Note that the difference between terms that are δ samples apart is

$$(n^2 - (n - \delta)^2)k = (2\delta n - \delta^2)k. \tag{B.20}$$

When $\delta = 1$, we have the difference between adjacent terms $(2n - 1)k$, which is an odd integer. This means that the sequence given by (B.19) has terms with alternating parity. Thus, the difference between an odd-indexed term and an even-index term is an odd integer, which cannot be D. Also, it can be inferred that if the sequence is equally-spaced by D, the set of equally-spaced values should have indices of the same parity. In this case, the terms with a difference of D should be 2δ samples apart, which gives a difference of

$$4(n + \delta)\delta k = lD, \tag{B.21}$$

where l is an odd integer. If D does not have a factor of 4, then there is no integer solution for n. Therefore, no equally-spaced pattern is formed for the case where D does not contain a factor of 4.

Consider the case where $\delta = 2$, i.e., terms of the same parity. We have

$$(n^2 - (n - 2)^2)k = (4n - 4)k. \tag{B.22}$$

Focus on the even-indexed terms. Letting $n = 2m$, the differences of the adjacent terms in this sub-sequence are

$$(8m - 4)k \tag{B.23}$$

and these terms can be expressed as

$$\left(\frac{4 + (8m - 4)}{2}(m)k\right) \bmod 2D = (4m^2 k) \bmod 2D. \tag{B.24}$$

Since

$$(4m^2 k) \bmod 2D = 4\left(m^2 k \bmod \frac{D}{2}\right), \tag{B.25}$$

the scaled sub-sequence formed by the even-indexed terms has a similar form to (B.19) under a modulus of $D/2$. The even-index of the the sequence $m^2 k \bmod D/2$ is again determined by the sequence

$$m^2 k \bmod \frac{D}{4}. \tag{B.26}$$

One could repeat this to the even-index terms of a sub-sequence until the modulus associated with the governing sequence is $D/2^l$, where l is a sufficiently large integer that $D/2^l$ does not contain a factor of 4. In this case, this sub-sequence and all sequences containing it are not equally-spaced, meaning that the even-indexed terms of (B.19) do not have equally-spaced values. As the result, the sequence itself does not contain equally-spaced values. This means that when D is even in Case II, $e_2[n]$ does not exhibit an equally-spaced pattern around n_0 that gives $e_1[n_0] \approx pM/(2D)$ where p is odd. Thus, these corresponding wandering events should happen at the fundamental wander rate.   □

Appendix C
Analysis for MASH 2-1

In this appendix, expressions for the internal states of a MASH 2-1 will be derived. For the MASH 2-1 shown in Fig. 5.14, the output of the EFM2 can be written as

$$y_1[n] = \frac{1}{M}\left(x[n] + d[n] - (e_1[n] - 2e_1[n-1] + e_1[n-2])\right). \tag{C.1}$$

Rearranging gives

$$x[n] + d[n] = My_1[n] + (e_1[n] - 2e_1[n-1] + e_1[n-2]). \tag{C.2}$$

Since $x[n] \in [0, M-1]$,

$$x[n] + d[n] = (My_1[n] + (e_1[n] - 2e_1[n-1] + e_1[n-2])) \bmod M \tag{C.3}$$

$$= (e_1[n] - e_1[n-1] - (e_1[n-1] - e_1[n-2])) \bmod M. \tag{C.4}$$

Therefore,

$$x[n] + d[n] = (e_1[n] - e_1[n-1] - (e_1[n-1] - e_1[n-2])) \bmod M, \tag{C.5}$$

$$x[n-1] + d[n-1] = (e_1[n-1] - e_1[n-2] - (e_1[n-2] - e_1[n-3])) \bmod M, \tag{C.6}$$

$$\vdots$$

$$x[0] + d[0] = (e_1[0] - e_1[-1] - (e_1[-1] - e_1[-2])) \bmod M. \tag{C.7}$$

D. Mai, M. P. Kennedy, *Wandering Spurs in MASH-based Fractional-N Frequency Synthesizers*, Analog Circuits and Signal Processing, https://doi.org/10.1007/978-3-030-91285-7

Summing gives

$$\left(\sum_{k=0}^{n}(x[k]+d[k])\right) \bmod M = (e_1[n] - e_1[-1] - (e_1[n-1] - e_1[-2])) \bmod M.$$
(C.8)

Therefore, equivalently,

$$\left((e_1[-1] - e_1[-2]) + \sum_{k=0}^{n}(x[k]+d[k])\right) \bmod M = (e_1[n] - e_1[n-1]) \bmod M.$$
(C.9)

If first-order dither

$$d[n] = d'[n] - d'[n-1]$$
(C.10)

is applied, where $d'[n]$ is a uniformly distributed between $[0, M-1]$, then we can further write

$$(e_1[n] - e_1[n-1]) \bmod M$$

$$= \left((e_1[-1] - e_1[-2]) - d'[-1] + d'[n] + \sum_{k=0}^{n}x[k]\right) \bmod M$$
(C.11)

The initial value $d'[-1]$ contributes as a constant and, in the following analysis, it is assumed that $d'[-1] = 0$.

The summation result indicates that the difference of $e_1[n]$, i.e.,

$$\nabla e_1[n] = e_1[n] - e_1[n-1]$$
(C.12)

is related to the running sum of the input and the dither. A MASH 1-1-1 with the same first-order shaped dither injected at its input has the quantization error $e_1[n]$ that can be expressed as

$$e_1[n] = (e_1[n]) \bmod M = \left(s_1[0] + d'[n] + \sum_{k=0}^{n}x[k]\right) \bmod M.$$
(C.13)

An equivalence can therefore be found between $\nabla e_1[n]$ of a MASH 2-1 and $e_1[n]$ of a MASH 1-1-1 with first-order shaped dither at its first stage. The *difference* between the initial condition $e_1[-1]$ and $e_1[-2]$ in the EFM2 is equivalent to the initial condition $s_1[0]$ of the MASH 1-1-1 with first-order shaped dither.

Since

$$(e_1[n] - e_1[n-1]) \bmod M = \left((e_1[-1] - e_1[-2]) + d'[n] + \sum_{k=0}^{n} x[k] \right) \bmod M,$$
$$(C.14)$$

$(e_1[n-1] - e_1[n-2]) \bmod M$

$$= \left((e_1[-1] - e_1[-2]) + d'[n-1] + \sum_{k=0}^{n-1} x[k] \right) \bmod M,$$
$$(C.15)$$

$$\vdots$$

$$(e_1[0] - e_1[-1]) \bmod M = \left((e_1[-1] - e_1[-2]) + d'[0] + \sum_{k=0}^{0} x[k] \right) \bmod M$$
$$(C.16)$$

Accumulation gives

$$e_1[n] = \left(e_1[-1] + (n+1)(e_1[-1] - e_1[-2]) + \sum_{k=0}^{n} d'[k] + \sum_{p=0}^{n} \sum_{k=0}^{p} x[k] \right) \bmod M$$
$$(C.17)$$

$$\approx \left(e_1[-1] + (n+1)(e_1[-1] - e_1[-2] + \mathbf{E}(d'[k])) + \sum_{p=0}^{n} \sum_{k=0}^{p} x[k] \right) \bmod M.$$
$$(C.18)$$

For a MASH 1-1-1 with first-order shaped dither injected at the input of the second stage EFM1, the second stage quantization error $e_2[n]$ is

$$e_2[n] = \left(s_2[0] + \sum_{p=0}^{n} (e_1[p] + d[p]) \right) \bmod M \qquad (C.19)$$

$$= \left(s_2[0] + (n+1)s_1[0] + \sum_{p=0}^{n} d[p] + \sum_{p=0}^{n} \sum_{k=0}^{p} x[k] \right) \bmod M \qquad (C.20)$$

$$\approx \left(s_2[0] + (n+1)(s_1[0] + \mathbf{E}(d[n])) + \sum_{p=0}^{n} \sum_{k=0}^{p} x[k] \right) \bmod M. \qquad (C.21)$$

Comparison between (C.18) and (C.21) shows the equivalence between $e_1[n]$ of MASH 2-1 and the second stage quantization error $e_2[n]$ of the first-order shaped dithered MASH 1-1-1.

Therefore, in conclusion,

$$s_{1,MASH111}[0] \equiv (e_1[-1] - e_1[-2]), \tag{C.22}$$

$$s_{2,MASH111}[0] \equiv e_1[-1]. \tag{C.23}$$

The expression for the accumulated quantization error can be found in a similar way. First, we have

$$e_{acc}[n] = \sum_{m=0}^{n-1} \left(y[m] - \frac{X}{M} \right) \tag{C.24}$$

$$= \sum_{m=0}^{n-1} y_1[m] + (y_2[n-1] - y_2[n-2]). \tag{C.25}$$

The summation of y_1 gives

$$\sum_{m=0}^{n-1} y_1[m] = \frac{1}{M} \left(nX + \sum_{m=0}^{n-1} d[m] - e_1[n-1] + e_1[n-2] + e_1[-1] - e_1[-2] \right). \tag{C.26}$$

The EFM1 in the MASH 2-1 has an output of

$$y_2[n] = \frac{1}{M} (e_1[n] - (e_2[n] - e_2[n-1])) \tag{C.27}$$

and therefore

$$y_2[n-1] - y_2[n-2] =$$
$$\frac{1}{M} (e_1[n-1] - e_1[n-2] - (e_2[n-1] - 2e_2[n-2] + e_2[n-3])). \tag{C.28}$$

Consider the first-order dither from (C.10); then we have

$$e_{acc}[n] = \sum_{m=0}^{n-1} y_1[m] + (y_2[n-1] - y_2[n-2]) \tag{C.29}$$

$$= \frac{1}{M} \left(nX + d'[n-1] - d'[-1] + (e_1[-1] - e_1[-2]) - \nabla^2 e_2[n-1] \right), \tag{C.30}$$

where

$$\nabla^2 e_2[n-1] = e_2[n-1] - 2e_2[n-2] + e_2[n-3].$$ (C.31)

Compare (C.30) with the corresponding expression for the MASH 1-1-1 (4.10) which is reproduced here

$$e_{acc}[n] = \frac{-1}{M}\left(\nabla^2 e_3[n-1] - s_1[0] - d[n-1]\right),$$ (C.32)

the equivalence

$$s_{1,MASH111}[0] \equiv -d'[-1] + (e_1[-1] - e_1[-2])$$ (C.33)

exists and, if $d'[-1] = 0$ is assumed, then we have

$$s_{1,MASH111}[0] \equiv (e_1[-1] - e_1[-2]),$$ (C.34)

which agrees with the prior result in (C.22).

In conclusion, the internal state $e_1[n]$ of the MASH 2-1 contains the parabolic structure for Case I and Case II inputs and contains an apparent linear pattern in Case III. In Case I and Case II, the dominant $\nabla^2 e_2$ component in the accumulated quantization error of the MASH 2-1 does not show wandering spurs unless it encounters nonlinearity. In Case III, the linear pattern $e_1[n]$ leads to a $\nabla^2 e_2[n]$ sequence that exhibits the underlying wandering spur pattern directly.

References

1. B. Miller, R.J. Conley, A multiple modulator fractional divider. IEEE Trans. Instrum. Measur. **40**(3), 578–583 (1991)
2. T.A.D. Riley, M.A. Copeland, T.A. Kwasniewski, Delta-Sigma modulation in fractional-N frequency synthesis. IEEE J. Solid-State Circ. **28**(5), 553–559 (1993)
3. ADF4356: 6.8 GHz Wideband Synthesizer with Integrated VCO , Analog Devices, Accessed: Apr. 16, 2020. [Online]. Available: https://www.analog.com/media/en/technical-documentation/data-sheets/ADF4356.pdf
4. M. Perrott, M. Trott, C. Sodini, A modeling approach for Σ-Δ fractional-N frequency synthesizers allowing straightforward noise analysis. IEEE J. Solid-State Circ. **37**(8), 1028–1038 (2002)
5. H. Arora, N. Klemmer, J.C. Morizio, P.D. Wolf, Enhanced phase noise modeling of fractional-N frequency synthesizers. IEEE Trans. Circ. Syst. I Regul. Pap. **52**(2), 379–395 (2005)
6. H. Hedayati, B. Bakkaloglu, W. Khalil, Closed-loop nonlinear modeling of wideband $\Sigma\Delta$ fractional-N frequency synthesizers. IEEE Trans. Microw. Theory Tech. **54**(10), 3654–3663 (2006)
7. P. Brennan, H. Wang, D. Jiang, P. Radmore, A new mechanism producing discrete spurious components in fractional-N frequency synthesizers. IEEE Trans. Circ. Syst. I Regul. Pap. **55**(5), 1279–1288 (2008)
8. A. Swaminathan, A. Panigada, E. Masry, I. Galton, A digital requantizer with shaped requantization noise that remains well behaved after nonlinear distortion. IEEE Trans. Signal Process. **55**(11), 5382–5394 (2007)
9. E. Familier, I. Galton, A fundamental limitation of DC-free quantization noise with respect to nonlinearity-induced spurious tones. IEEE Trans. Signal Process. **61**(16), 4172–4180 (2013)
10. Z. Li, H. Mo, M.P. Kennedy, Comparative spur performance of a fractional-N frequency synthesizer with a nested MASH-SQ3 divider controller in the presence of memoryless piecewise-linear and polynomial nonlinearities, in *25th IET Irish Signals Systems Conference 2014 and 2014 China-Ireland International Conference on Information and Communications Technologies (ISSC 2014/CIICT 2014)*. (IEEE, New York, 2014), pp. 374–379
11. D. Mai, M.P. Kennedy, A design method for nested MASH SQ hybrid divider controllers for fractional-N frequency synthesizers. IEEE Trans. Circ. Syst. I Regul. Pap. **65**(10), 3279–3290 (2018)
12. E. Perez Gonzalez, J. Reiss, Idle tone behavior in Sigma Delta modulation, in *122nd Audio Engineering Society (AES) Conference*, May 2007

13. D. Mai, H. Mo, M.P. Kennedy, Observations and analysis of wandering spurs in MASH-based fractional-N frequency synthesizers. IEEE Trans. Circ. Syst. II: Express Briefs **65**(5), 662–666 (2018)

14. L. Grimaldi, D. Cherniak, W. Grollitsch, R. Nonis, Analysis of spurs impact in PLL-based FMCW radar systems, in *2020 IEEE International Symposium on Circuits and Systems (ISCAS)*, October (IEEE, New York, 2020)

15. D. Mai, M.P. Kennedy, Analysis of wandering spur patterns in a fractional-N frequency synthesizer with a MASH-based divider controller. IEEE Trans. Circ. Syst. I Regul. Pap. **67**(3), 729–742 (2020)

16. M. Perrott, *CppSim Reference Manual*, 2014 [Online]. Available: https://www.cppsim.com/Manuals/cppsimdoc.pdf

17. M.P. Kennedy, G. Hu, V.S. Sadeghi, Observations concerning noise floor and spurs caused by static charge pump mismatch in fractional-N frequency synthesizers, in *Irish Signals Systems Conference 2014 and 2014 China-Ireland Internatio nal Conference on Information and Communications Technologies (ISSC 2014/CIICT 2014). 25th IET*, June (IEEE, New York, 2014), pp. 24–29

18. M.P. Kennedy, V.S. Sadeghi, Observations concerning PFD/CP operating point offset strategies for combatting static charge pump mismatch in fractional-N frequency synthesizers with digital delta-sigma modulators. Nonlinear Theory Appl. **5**(3), 349–364 (2014)

19. S. Bou Sleiman, J.G. Atallah, S. Rodriguez, A. Rusu, M. Ismail, Optimal $\Sigma\Delta$ modulator architectures for fractional-N frequency synthesis. IEEE Trans. Very Large Scale Integr. (VLSI) Syst. **18**(2), 194–200 (2010)

20. P. Kartaschoff, *Frequency and Time*. Monographs in Physical Measurement (Academic, New York, 1978)

21. *1139–1999 - IEEE Standard Definitions of Physical Quantities for Fundamental Frequency and Time Metrology-Random Instabilities*. New York, USA: IEEE, 1999. [Online]. Available: http://www.photonics.umbc.edu/Menyuk/Phase-Noise/Vig_IEEE_Standard_1139-1999.pdf

22. A.V. Oppenheim, R.W. Schafer, J.R. Buck, *Discrete-Time Signal Processing* (Prentice-Hall, Upper Saddle River, NJ, 1999)

23. M.H. Hayes, *Statistical Digital Signal Processing and Modeling* (Wiley, New York, 1996)

24. W. Egan, *Advanced Frequency Synthesis* (Wiley, New York, 2011)

25. K.J. Wang, A. Swaminathan, I. Galton, Spurious tone suppression techniques applied to a wide bandwidth 2.4GHz fractional-N PLL. IEEE J. Solid-State Circ. **43**(12), 2787–2797 (2008)

26. E. Familier, I. Galton, Second and third-order noise shaping digital quantizers for low phase noise and nonlinearity-induced spurious tones in fractional-N PLLs. IEEE Trans. Circ. Syst. I Regul. Pap. **63**(6), 836–847 (2016)

27. B. Razavi, An alternative analysis of noise folding in fractional-N synthesizers, in *2018 IEEE International Symposium on Circuits and Systems (ISCAS)*, May (IEEE, New York, 2018), pp. 1–4

28. M.P. Kennedy, H. Mo, G. Hu, Comparison of a feed-forward phase domain model and a time domain behavioral model for predicting mismatch-related noise floor and spurs in fractional-N frequency synthesizers, in *2015 26th Irish Signals and Systems Conference (ISSC)*, June (IEEE, New York, 2015), pp. 1–6

29. D. Mai, H. Mo, M.P. Kennedy, Observations of the differences between closed-loop behavioral and feed-forward model simulations of fractional-N frequency synthesizers, in *2017 28th Irish Signals and Systems Conference (ISSC)*, June (IEEE, New York, 2017), pp. 1–6

30. D. Mai, A. Dahlan, M.P. Kennedy, MASH DDSM-induced spurs in a fractional-N frequency synthesizer, in *2019 26th IEEE International Conference on Electronics, Circuits and Systems (ICECS)*, November (IEEE, New York, 2019), pp. 13–16

31. D. Banerjee, *PLL Performance, Simulation and Design* (Dog Ear Publishing, Indianapolis, IN, 2017)

32. V. Mazzaro, M.P. Kennedy, Observations concerning "horn spurs" in a MASH-based fractional-N CP-PLL, in *2019 26th IEEE International Conference on Electronics, Circuits and Systems (ICECS)*, November (IEEE, New York, 2019), pp. 17–20

33. *R&S® FSW Real-TimeSpectrum Application and MSRT Operating Mode User Manual*, Rohde & Schwarz. [Online]. Available: https://scdn.rohde-schwarz.com/ur/pws/dl_downloads/pdm/cl_manuals/user_manual/1175_6484_01/FSW_Realtime_UserManual_en_18.pdf

34. D. Mai, X. Li, M.P. Kennedy, Experimental confirmation of wandering spurs in a commercial fractional-N frequency synthesizer with a MASH 1-1-1 divider controller, in *2019 30th Irish Signals and Systems Conference (ISSC)*, June (IEEE, New York, 2019), pp. 1–6

35. *ADF 4159 Direct Modulation/Fast Waveform Generating, 13 GHz, Fractional-N Frequency Synthesizer*, Analog Devices, Accessed: Dec. 17, 2019. [Online]. Available: https://www.analog.com/media/en/technical-documentation/data-sheets/ADF4159.pdf

36. Analog Devices, *EV-ADF4159EB1Z/EV-ADF4159EB3Z User Guide*, 2015. [Online]. Available: https://www.analog.com/media/en/technical-documentation/evaluation-documentation/UG-383.pdf

37. A. Swaminathan, K. Wang, I. Galton, A wide-bandwidth 2.4 GHz ISM band fractional-N PLL with adaptive phase noise cancellation. IEEE J. Solid-State Circ. **42**(12), 2639–2650 (2007)

38. J. Song, I.C. Park, Spur-free MASH delta-sigma modulation. IEEE Trans. Circ. Syst. I: Regul. Pap. **57**(9), 2426–2437 (2010)

39. K. Hosseini, M.P. Kennedy, Maximum sequence length MASH digital delta–sigma modulators. IEEE Trans. Circ. Syst. I Regul. Pap. **54**(12), 2628–2638 (2007)

40. B. Fitzgibbon, S. Pamarti, M.P. Kennedy, A spur-free MASH DDSM with high-order filtered dither. IEEE Trans. Circ. Syst. II Express Briefs **58**(9), 585–589 (2011)

41. D. Yang, F. Dai, W. Ni, S. Yin, R. Jaeger, Delta-sigma modulation for direct digital frequency synthesis. IEEE Trans. Very Large Scale Integr. Syst. **17**(6), 793–802 (2009)

42. D. Mai, M.P. Kennedy, Influence of initial condition on wandering spur pattern in a MASH-based fractional-N frequency synthesizer. IEEE Trans. Circ. Syst. II Express Briefs **67**(12), 2968–2972 (2020)

43. S. Pamarti, I. Galton, LSB dithering in MASH Delta-Sigma D/A converters. IEEE Trans. Circ. Syst. I Regul. Pap. **54**(4), 779–790 (2007)

44. H. Mo, M.P. Kennedy, Apparatus for reducing wandering spurs in a fractional-N frequency synthesizer, Patent US 10,541,707 (2018)

45. V.R. Gonzalez-Diaz, M.A. Garcia-Andrade, G.E. Flores-Verdad, F. Maloberti, Efficient dithering in MASH sigma-delta modulators for fractional frequency synthesizers. IEEE Trans. Circ. Syst. I Regul. Pap. **57**(9), 2394–2403 (2010)

46. H. Huh, Y. Koo, K.-Y. Lee, Y. Ok, S. Lee, D. Kwon, J. Lee, J. Park, K. Lee, D.-K. Jeong, W. Kim, Comparison frequency doubling and charge pump matching techniques for dual-band $\Delta\Sigma$ fractional-N frequency synthesizer. IEEE J. Solid-State Circ. **40**(11), 2228–2236 (2005)

47. M.P. Kennedy, Y. Donnelly, J. Breslin, S. Tulisi, S. Patil, C. Curtin, S. Brookes, B. Shelly, P. Griffin, M. Keaveney, 16.9 4.48GHz 0.18 μm SiGe BiCMOS exact-frequency fractional-N frequency synthesizer with spurious-tone suppression yielding a −80 dBc in-band fractional spur, in *2019 IEEE International Solid-State Circuits Conference - (ISSCC)*, February 2019

48. Y. Donnelly, M. Keaveney, M.P. Kennedy, J. Breslin, S. Tulisi, S. Patil, C. Curtin, S. Brookes, B. Shelly, P. Griffin, 4.48-GHz fractional-N frequency synthesizer with spurious-tone suppression via probability mass redistribution. IEEE Solid-State Circ. Lett. **2**(11), 264–267 (2019)

49. Y. Donnelly, H. Mo, M.P. Kennedy, High-speed nested cascaded MASH digital delta-sigma modulator-based divider controller, in *2018 IEEE International Symposium on Circuits and Systems (ISCAS)*, May (IEEE, New York, 2018)

Index

Printed in the United States
by Baker & Taylor Publisher Services